Roadmap: Restoration of the Salton Sea Ecosystem

ANTONIOS VALAMONTES

"A Personal Journey to Salton Sea: My Visions of the Lake Spirit"

In the ethereal realm of dreams, guided by the wisdom of the Hualapai Medicine Man, I embarked on a profound and transformative journey—one that interwove the enigmatic allure of the Salton Sea with the depths of my own consciousness.

In the tapestry of my dreams, the Lake Spirit emerged as a sentinel of ancient knowledge and ethereal guidance. The Hualapai Medicine Man, a revered figure from the realm of dreams, bestowed upon me a spiritual quest—an odyssey that transcended the boundaries of the waking world. Through his guidance, I ventured into the heart of the Salton Sea's mystique, unraveling its intricate narratives of resilience and transformation.

As the Medicine Man's words echoed in the chambers of my mind, I found myself transported to the shores of the Salton Sea. The vast expanse of water stretched before me, shimmering with an otherworldly luminescence. Yet, beneath its surface, I sensed the echoes of a poignant tale—one of a lake that had witnessed the ebb and flow of civilizations, the rise and fall of hope, and the tenacity of life amid adversity.

Guided by the Lake Spirit's whispers, I delved deeper into the waters of introspection. Each ripple of the lake seemed to carry the stories of countless generations, echoing the struggles and aspirations of those who had come before. The lake's narrative unfolded before me—a saga of birth, decay, and rebirth, mirroring the cycles of existence itself.

Through the Medicine Man's teachings, I learned to decipher the language of the wind and the currents and to read the reflections on the water's surface as a tapestry of truths. The lake, once a symbol of serenity, had been marked by ecological challenges—a decline in water quality, the loss of habitats, and the plight of its inhabitants. Yet, the Lake Spirit's essence remained resilient, a testament to the enduring spirit of nature's creations.

As dawn broke across the horizon, the Medicine Man's presence lingered, a reminder of the profound connection between the realms of dreams and reality. Just as he had guided me in my visions, so too did he inspire me to embrace a journey of restoration, both for the lake and the depths of my being.

The Salton Sea, a canvas of dreams and realities, reflected the synergy between the metaphysical and the tangible. Its story resonated with the echoes of my own quest—an exploration of purpose, meaning, and the intricate interplay between human aspirations and the natural world.

In this profound union of dreams and reality, the Lake Spirit's guidance illuminated a path forward—a path that entwined personal growth with environmental stewardship. The journey I undertook became a testament to the intricate dance of interconnectedness—a dance that transcends boundaries and invites us to become guardians of the Earth's mysteries.

And so, the visions of the Lake Spirit, guided by the Hualapai Medicine Man, remain etched in the tapestry of my journey—a journey that converges with the pulsating heart of the Salton Sea, beckoning for restoration, renewal, and the harmonious unity of all beings.

Introduction

The Salton Sea, located in Southern California, has long been facing ecological challenges due to declining water levels, high salinity, and environmental degradation. This roadmap examines the potential impact of introducing one million liters of fresh water daily into the Salton Sea to restore its fragile ecosystem. Through a multidisciplinary approach that combines hydrology, ecology, and environmental management, this research aims to shed light on the feasibility, benefits, and implications of this intervention to revitalize the lake's environment.

Adding a million liters of fresh water per day to the Salton Sea can have significant effects on the lake's overall salinity, ecosystem, and water quality. However, the exact impact would depend on various factors, including the current conditions of the Salton Sea, the distribution of the added water, and the lake's hydrology.

Here are some potential effects:

1. **Salinity Reduction:** The added freshwater would dilute the salinity of the Salton Sea, which is essential for maintaining a balanced ecosystem. High salinity levels can harm aquatic life and contribute to water quality issues.

2. **Aquatic Ecosystem:** The lower salinity resulting from adding freshwater could create a more hospitable environment for certain aquatic species. This might encourage the growth of fish, birds, and other wildlife better suited to lower salinity conditions.

3. **Water Quality Improvement:** The influx of fresh water can help improve water quality by reducing pollutants and contaminants that might be present in the lake.

4. **Habitat Expansion:** Lower salinity levels could potentially expand habitats for aquatic plants and animals, supporting a more diverse ecosystem.

5. Evaporation Mitigation: The added freshwater might also reduce the evaporation rate, helping stabilize the lake's water levels.

6. Economic Impact: A healthier ecosystem and improved water quality could positively impact local communities by supporting recreational activities and tourism.

7. Complex Ecosystem Dynamics: While diluting the salinity could benefit some species, it could also disrupt the balance of the existing ecosystem. The changes might negatively affect some species, while others could thrive.

8. Long-Term Management: The addition of fresh water should be part of a comprehensive long-term management strategy that considers factors such as inflow, outflow, water quality, and the overall health of the ecosystem.

It's important to note that any large-scale intervention in an ecosystem should be carefully planned and studied to understand potential consequences fully. Consulting with hydrologists, ecologists, and environmental experts would be essential to assess the potential benefits and risks of adding fresh water to the Salton Sea.

Chapter 1: Background and Significance

1.1 Introduction

The Salton Sea, nestled within the arid expanse of Southern California, has been grappling with ecological degradation for an extended period. A complex interplay of factors, ranging from dwindling water levels and elevated salinity to restricted inflow from neighboring sources, has precipitated this environmental decline. Once a vibrant and bustling ecosystem cherished both for its natural beauty and recreational allure, the lake's downward trajectory has precipitated a slew of ecological challenges. Habitat loss, dwindling aquatic populations, and a concerning decline in water quality have emerged as prominent issues. Given its status as the largest lake in California, the Salton Sea's deterioration carries far-reaching implications, spanning regional biodiversity, water resource management, and the well-being of local communities.

1.2 Historical Context

The narrative of the Salton Sea unfolds as a tapestry woven with threads of human intervention and nature's response, a story intricately connected to the ebb and flow of time. From accidental flooding that birthed its waters to the nuanced dance of water management practices, the lake's history is an enigmatic journey etched by the hands of fate.

Within its chapters, the narrative ebbs and flows, tracing the contours of vitality and vulnerability that have sculpted its essence. Against this backdrop, the stage is set for the evaluation of contemporary restoration efforts. This inquiry navigates the currents of history to unearth the essence of the Salton Sea's narrative.

1.2.1 Accidental Genesis: The Birth of the Salton Sea

The captivating narrative of the Salton Sea's origins unfolds like a tale of chance and fate, tracing its roots to a remarkable series of events that transpired in the early 20th century. This chapter of history reveals a fascinating convergence of human endeavors and the forces of nature, resulting in the creation of the enigmatic ecosystem known as the Salton Sea.

The saga begins with a burst levee, a rupture in the defenses that separated the Colorado River from the arid expanse of Southern California. As the river's waters surged beyond their confines, they carved a new path, inundating the desert landscape and forming an unexpected basin that would become the foundation for the Salton Sea. It was a moment of sheer happenstance where the hand of nature brushed aside human-made boundaries, reshaping the land in its wake.

Amidst this unfolding drama, human ambition and innovation found their place in the narrative. Recognizing the potential for harnessing this newly formed body of water, local communities and developers sought to transform the Salton Sea into a hub of recreation and prosperity. Marinas, resorts, and communities emerged along its shores, driven by the allure of a waterfront paradise in the heart of the desert.

However, the delicate balance between human intentions and nature's whims soon revealed itself. As decades passed, the Salton Sea faced its share of challenges, including fluctuating water levels, escalating salinity, and ecological imbalances. The very forces that had once birthed the sea now posed threats to its sustainability, leading to a complex interplay between human intervention and ecological resilience.

In the accidental genesis of the Salton Sea, a story of contrasts emerges – of how the accidental and the intentional, the natural and the human-made, converged to create a landscape of both wonder and complexity. As we delve further into this narrative, we unravel the intricate threads that weave together the past, present, and potential future of the Salton Sea – a tale that reminds us of the intricate dance between our aspirations and the forces of the world around us.

1.2.2 A Symphony of Shifting Waters: The Ever-Evolving Fate of the Salton Sea

As time unfurls its tapestry, the story of the Salton Sea transforms into a symphony characterized by the ebb and flow of shifting waters. In this mesmerizing composition, the orchestra is composed of water management practices, agricultural runoff, and the intricate choreography of hydrological dynamics. Through the passages of triumphs and tribulations, the lake's journey emerges as a captivating dance between human agency and the harmonious cadence of the natural world, with each movement etching its imprint on the canvas of its destiny.

The score begins with the notes on water management practices, a delicate arrangement of policies and strategies crafted by human hands. These efforts aimed to navigate the balance between utilizing the sea's resources and safeguarding its fragile ecosystem. The lake's role as an agricultural reservoir and a source of freshwater supply brought forth a series of orchestrated actions, shaping its trajectory against the backdrop of evolving societal needs.

As the symphony progresses, the echoes of agricultural runoff reverberate throughout the narrative. The runoff, laden with nutrients and sediments, tells a tale of human endeavors colliding with the environment. The dance between irrigation and drainage systems, coupled with the residues of agricultural practices, composed a complex melody that played a pivotal role in shaping the sea's composition, for better or worse.

The composition's crescendo arrives with the whispers of changing hydrological dynamics, the undercurrents that define the lake's interactions with surrounding watersheds. The fluctuations in water inflows from the Colorado River, the rise and fall of water levels, and the symphony of evaporation all merge to create a harmonious yet unpredictable melody. The result is a tapestry of interconnected rhythms that define the Salton Sea's life story.

In this symphony of shifting waters, the Salton Sea's journey unfolds as a testament to the intricate interplay between human stewardship and the forces of nature. Just as a symphony evolves with every conductor's interpretation, the lake's destiny is composed through the choices, interventions, and respect accorded by those who have encountered its shores. And as the symphony continues, the tale of the Salton Sea remains an ongoing performance that weaves together the threads of humanity's aspirations with the eternal rhythms of the Earth.

1.2.3 Dynamics of Vulnerability: Navigating the Delicate Balance of the Salton Sea

The annals of the Salton Sea's history unfold as a narrative intricately woven with threads of vulnerability. In this tapestry, the harmonious dance of water inflow and outflow orchestrates the very existence of the lake. Within these pages, the reader encounters a story of juxtaposition—moments of abundance seamlessly merging with intervals of scarcity, each revealing the symphony of interplay that intricately defines the essence of the Salton Sea. It is a history that embodies resilience, a testament to nature's ability to withstand environmental challenges, and a testament to human determination in the pursuit of equilibrium.

From the inception of the lake, the dynamics of its vulnerability were set in motion. The delicate equilibrium between water sources and the relentless grasp of evaporation constantly tilts the scales. The chapters of abundance unfold with the rush of inflows from the

Colorado River, nourishing the sea and heralding moments of bounty. Yet, with every chapter of Plenty comes the refrain of vulnerability as the specter of scarcity looms with evaporation's persistent call.

In the symphony of this vulnerability, the Salton Sea's history resonates as a testament to the lake's remarkable adaptability and resilience. The waters have seen fluctuations in volume and shifts in salinity, yet life endures. Avian species find haven on its shores, adapting to the changing tides of habitat availability. Local communities, too, have navigated the dance of nature's whims, finding ways to harness the sea's resources while acknowledging the constraints of its delicate balance.

As history marches forward, the lake's legacy stands as a testament to the triumph of resilience over adversity. It underscores the capacity of both nature and humanity to adapt and endure, even in the face of the vulnerability inherent in the Salton Sea's existence. The lake's story is a reminder that while vulnerability may be woven into its fabric, the strength of its resilience is equally woven into its narrative. Through the ebb and flow of history, the Salton Sea remains a living embodiment of equilibrium's enduring pursuit, a lesson that vulnerability and resilience are two sides of the same coin in the ever-evolving dance of nature.

1.2.4 Vistas of Restoration: Tracing the Threads of Past, Present, and Renewal

Salton Bay Yacht Club – Salton City, California
South of Palm Springs, between Indio and El Centro via Highway 99

In the tapestry of the present moment, the echoes of

history cast their subtle shadows on the canvas of restoration efforts. The narratives of yesteryears, both their challenges and triumphs, become integral voices in the ongoing conversations of the present. They become the guiding stars that illuminate the path toward restoring the delicate equilibrium between human aspiration and the ecological integrity of the Salton Sea.

Against the backdrop of history, the chapters that unfold in this narrative offer a profound exploration of the complexities that underlie the rejuvenation endeavors of today. The echoes of accidental genesis and the symphony of shifting waters have forged a narrative that sets the stage for restoration aspirations. The vulnerability woven into the very essence of the lake beckons us to tread thoughtfully as we seek to mend the delicate dance between human intervention and nature's rhythms.

Within this historical tapestry, each thread is an invitation to explore the multifaceted dimensions of the Salton Sea's story. The accidental genesis of the lake's creation whispers tales of serendipity and the interplay between human ambition and nature's unpredictability. The symphony of shifting waters reminds us of the ebb and flow of human influence and the rhythms of the natural world. The dynamics of vulnerability speak of the balance between abundance and scarcity, resilience and fragility.

Yet, amidst these threads, a beacon of hope shines brightly: the vistas of restoration. In the stories of the past, we find both the cautionary tales and the inspirations that guide our footsteps today. As we embark on the journey of restoration, we are called to honor the intricacies of the lake's history and heed the lessons it imparts. The chapters that unfold within these pages offer glimpses of the ongoing restoration efforts, rooted in a profound understanding of the past and fueled by a shared commitment to nurturing the Salton Sea back to its ecological vibrancy.

This narrative is an invitation to explore the interplay between past, present, and renewal, where threads of history converge to weave a tale that is at once a chronicle of challenges and a testament to human resilience. As the stories of accidental genesis, shifting waters, vulnerability, and restoration aspirations intertwine, a narrative of hope emerges—a story that celebrates the potential for collective action, the promise of renewal, and the enduring quest to restore the delicate harmony between humanity and the living ecosystem of the Salton Sea.

1.3 Ecological and Environmental Challenges: Unraveling the Complex Web

The challenges that converge upon the Salton Sea create a tapestry of complexity, interwoven with multifaceted threads that demand our attention. Among these challenges, the decline of water levels stands as a prominent concern, a consequence amplified by the confluence of reduced inflow from the Colorado River and regional water diversions. This unfortunate synergy has unveiled vast expanses of the lakebed, where once glistening waters danced.

Yet, this newfound exposure carries a darker tale—a tale woven with harmful dust particles laden with pollutants. These particles, awakened from their slumber by the caress of the wind, traverse the barren lakebed to embrace the skies. Their journey knows no boundaries, and their embrace carries the potential to cast a shadow of health risks upon the wings of wildlife and the shoulders of communities dwelling nearby.

Amidst this ecological intricacy, the haunting refrain of elevated salinity echoes. It reverberates through the water, shaping an environment that becomes inhospitable to many native species. With each rise in salinity, a symphony of challenges unfolds—a symphony that orchestrates declines in aquatic populations and alters the very composition of the lake's inhabitants. The Salton Sea, once teeming with diversity, finds its voice muted as the delicate balance of its ecosystem falters.

From the decline of water levels to the release of dust-laden pollutants and the specter of elevated salinity, each challenge has a voice to share—a voice that speaks of a complex ecosystem in peril. Yet, within the shadows, there is a glimmer of hope—a hope that the insights we gain through exploration and understanding will pave the way for solutions that breathe life back into this intricate tapestry of nature.

1.4 Significance of Restoration: A Ripple in the Web of Life

The call to restore the Salton Sea's delicate ecological equilibrium transcends the boundaries of its shores. This imperative resonates far beyond the surface waters, weaving a narrative that intertwines with interconnected landscapes and reverberates through the intricate fabric of our environment.

The struggles of this unique ecosystem are not confined to its watery borders. The loss of critical habitat, a consequence of the ecological unraveling, echoes the cries of displaced species. It sends ripples through the tapestry of regional biodiversity, where each thread of life is woven into a complex web of interdependence. The disruptions to avian migration patterns, once guided by the timeless rhythm of the lake, become a dissonant note in the symphony of nature.

Yet, the significance of restoration extends even further, weaving tendrils into the very fabric of human existence. Local communities, whose lives are intertwined with the ebb and flow of the Salton Sea, stand as witnesses to the consequences of its decline. The economic contributions that once flowed from the lake's vitality have ebbed, leaving behind a void that echoes in the economic pulse of the region. Recreation opportunities, once abundant, have dwindled, leaving communities yearning for the embrace of the lake's beauty.

In this dance of interconnectedness, the restoration of the Salton Sea holds the promise of renewal—a renewal that extends beyond the shores, beyond the waters, and into the hearts of those who call this landscape home. The threads of restoration, guided by the imperative of ecological balance, seek to mend the tapestry of life that has been frayed. They endeavor to harmonize the symphony of nature once more, allowing each note to resonate with purpose and meaning. As we delve into the chapters ahead, we uncover the layers of significance that restoration carries—a significance that ripples through ecosystems, communities, and the very essence of our shared planet.

1.5 Research Objectives

Amidst the arid canvas of the Salton Sea, this roadmap emerges as a compass, guiding an exploration into the realm of possibilities. With purposeful intent, it unfurls its banners, declaring its steadfast dedication to comprehensive assessment and illuminative inquiry. In the heart of this journey lies the formidable endeavor to introduce one million liters of fresh water daily, a concept that beckons curiosity and deliberation. The roadmap unfurls its banners, each fluttering with a unique objective, collectively weaving a tapestry of understanding and action.

1.5.1 Guiding the Voyage of Feasibility

At its very heart, this roadmap embarks on a purposeful journey through the uncharted territory of feasibility, guided by the unwavering compass of hydrological modeling. Through the lens of data-driven exploration, it aspires to decode the intricate symphony that plays out in the delicate dance of water balance, evaporation rates, and the ebb and flow of inflow patterns. With a resolute objective to cast light upon the plausibility of introducing fresh water, this roadmap aims to swing open the doors to hitherto untold possibilities.

Just as a ship's course is meticulously charted using the stars, this roadmap sets its course by navigating the data-driven constellations of hydrological information. By weaving together the threads of empirical evidence, it endeavors to unveil the pathway toward a restored equilibrium. In the ever-evolving tale of the Salton Sea's restoration, this chapter serves as a compass, guiding us toward the realm where dreams of transformation intersect with the reality of achievable change.

1.5.2 Unveiling the Radiance of Potential Benefits

Much like the sun's gentle rays illuminate the landscape, this roadmap seeks to shine a light on the potential benefits that lie within the realm of fresh water addition. Guided by the torch of meticulous ecological impact assessments, it embarks on a journey to unravel the intricate threads of ecosystems. From the delicate dance of aquatic species to the cascading effects of altered salinity, this roadmap's path is paved with the intention of revealing the positive reverberations that could blossom from this transformative intervention.

Just as the sun's warmth brings life and vitality to the world, the exploration of potential benefits carries the promise of breathing new life into the ecosystem. Through the lenses of ecological analysis, the roadmap seeks to unveil the hidden gems that await discovery—a myriad of possibilities that could offer respite to a landscape burdened by ecological challenges. As we tread along this path of exploration, we peer into the tapestry of nature's resilience and the potential for renewal that fresh water addition might bring.

1.5.3 Mapping Economic Vistas

Within this narrative, economics too claims its rightful place—a thread skillfully woven through the fabric of analysis and assessment. Through the lens of economic analyses, the roadmap turns its discerning gaze toward the horizon where local communities dwell. Here, the intention is to unearth the potential economic reverberations—those echoes of impact that resonate akin to the ripples coursing across the surface of the Salton Sea's waters.

Just as a cartographer meticulously traces the contours of a map, economic analysis plots the contours of potential benefits. It is a journey that ventures beyond numbers, delving into the lives of those who call the Salton Sea's vicinity home. With a compass aligned to the aspirations of communities and the potential for a thriving local economy, this chapter unfurls a canvas where financial value intertwines with the tapestry of restoration.

1.5.4 A Rallying Cry for Action

With empirical investigation as its lodestar and predictive modeling as its compass, this roadmap resounds with a clarion call for action. Its purpose is to illuminate the way forward, to equip decision-makers with insights forged through the crucible of data and contemplation. It stands as a beacon of possibility, a guiding star navigating the intricate currents of ecological interplay, human ambition, and the mystique of nature's choreography.

In this symphony of exploration and inquiry, the research objectives stand tall like unfurled banners—a choreographed dance of feasibility, a canvas brushed with potential benefits, a panorama of economic exploration, and above all, a rallying cry for proactive engagement. Each objective, a luminous star in its own celestial constellation, converges to chart a course toward a future where the Salton Sea's narrative might evolve into a tale of renaissance and revival.

Chapter 2: Literature Review

2.1 Salton Sea Ecosystem and Challenges

Nestled in the heart of Southern California's arid landscape, the Salton Sea stands as a testament to the intricate dance between human intervention and nature's delicate balance. Born from accidental flooding during early 20th-century engineering endeavors, this once-unplanned reservoir emerged as a unique and enigmatic feature in the region's topography. However, the passage of time has cast its shadow on ecological challenges upon this saline, closed-basin lake, presenting a complex puzzle of interactions that demand thoughtful consideration.

The origin story of the Salton Sea is one of accidental genesis, a tale of ambitious engineering projects that inadvertently gave birth to this water body. An intricate interplay of human activity and geological dynamics culminated in its formation, offering a glimpse into the unintended consequences that can unfold when human ambition meets the unpredictable forces of nature.

Yet, as the years have unfolded, the Salton Sea has encountered a host of ecological challenges that mirror the intricacies of its origins. The lake's unique geology and hydrology have woven a narrative of their own, weaving a tapestry of complexities that continue to shape its destiny. Amidst the shimmering expanse of water, rising salinity levels have emerged as a silent protagonist, driven by the relentless dance of evaporation that marks arid landscapes.

Compounding this challenge is the dwindling embrace of inflows from the once-mighty Colorado River, a vital lifeline that once nourished the Salton Sea's existence. As the river's offerings diminished, the lake found itself grappling with an insidious salinity stress that permeated its waters. An escalating battle between evaporative forces and dwindling inflows set the stage for an ecological struggle that reverberated through the lake's ecosystem.

It is within this intricate framework that the Salton Sea's aquatic inhabitants found themselves on the frontlines of adversity. The mounting salinity stress took its toll, weaving a tale of disrupted equilibrium that reverberated across the ecosystem's fabric. Biodiversity, once celebrated in the vibrant tapestry of species that called the lake home, found itself under siege. Aquatic habitats, once flourishing nurseries of life, bore witness to the erosion of water quality that eroded their vitality.

In the heart of this delicate balance, the Salton Sea's story unfolds—a saga of ecological intricacies, a symphony of challenges, and a call to action. As we delve into the complexities of this ecosystem and its trials, we embark on a journey to understand not only its past but also to shape its future—a future where restoration and rejuvenation can offer a fresh chapter of hope to a landscape longing for renewal.

2.2 The Role of Salinity in Ecosystem Degradation

Within the intricate web of the Salton Sea's ecosystem, the specter of salinity emerges as a formidable force that orchestrates a symphony of ecological challenges. This section unveils the critical role that elevated salinity levels play in the ongoing degradation of the lake's delicate balance—a role that weaves a narrative of disrupted harmony and imperiled habitats.

In the azure expanse of the Salton Sea, the consequences of heightened salinity levels ripple far beyond their mere numerical representation. Like an unseen hand, high salinity levels extend their influence, shaping the aquatic landscape in profound ways. As the salinity index rises, a cascade of challenges begins, suffusing the waters with implications that extend beyond the surface.

One of the most dramatic impacts of escalating salinity is the diminishment of dissolved oxygen levels—an essential component for the survival of aquatic species. The once-bountiful lake transforms into an inhospitable realm for many native inhabitants, forcing them to navigate a precarious existence amidst the diminishing oxygen supply. As salinity intensifies, the veil of oxygen thins, creating a barrier that suffocates aquatic life and leaves them grappling for survival.

The intricate dance between salinity and essential nutrients further compounds the challenges. Elevated salinity creates a hostile environment for essential nutrients that form the building blocks of aquatic sustenance. As the saline quotient increases, the availability of these vital nutrients diminishes, disrupting the delicate balance of the food chain. A once-thriving ecosystem finds itself ensnared in a web of nutrient imbalances, cascading through trophic levels and distorting the symphony of life.

The rise of salinity also lends itself to a sinister accomplice—algal blooms. These blooms, triggered by the convergence of heightened salinity and excess nutrients, unleash a wave of consequences that ripple across the lake's expanse. Algae, once benign components of the aquatic tapestry, proliferate uncontrollably under these conditions. Their rapid multiplication transforms the water's hue, casting a sickly shadow that disrupts the natural aesthetics and further imperils the lake's inhabitants.

The repercussions of high salinity levels extend far beyond their numeric representation, reaching into the very heart of the Salton Sea's intricate web of life. As salinity rises, the stage is set for a series of interconnected challenges that render the once-vibrant ecosystem vulnerable to disruption. Dissolved oxygen, nutrient availability, and algal blooms converge in a complex dance that shapes the lake's destiny, illuminating the urgency of addressing salinity as a pivotal step towards restoration and rejuvenation.

2.3 Previous Restoration Efforts and Their Outcomes

In the annals of the Salton Sea's history, a tapestry of restoration attempts unfurls, each endeavoring to mend the frayed threads of the ecosystem's vitality. This section delves into the chronicles of these past endeavors, tracing their paths from inception to outcome, and shedding light on the lessons etched in their wake.

Amidst the backdrop of the Salton Sea's ecological woes, a series of attempts have been woven into the narrative with the aim of resuscitating the fading vitality of the lake. From the introduction of non-native fish species to the redirection of water from agricultural runoff, a medley of strategies has been set into motion, each bearing the promise of rejuvenation. However, the intricate nature of the ecosystem's interactions and the labyrinthine pathways of regional water management have often cast shadows on the outcomes of these endeavors.

The mosaic of restoration efforts, while fueled by good intentions, has often yielded outcomes that fall shy of the envisioned renaissance. The introduction of non-native fish species, intended to recalibrate the aquatic balance, inadvertently set off ripples of unintended consequences. As the alien inhabitants interacted with native species, intricate ecological interplays were disrupted, and the promised equilibrium remained elusive.

Diverting water from agricultural runoff—another arrow in the quiver of restoration—proved a solution encased in complexities. As the currents of diverted water flowed into the Salton Sea, they brought with them a tapestry of nutrients and sediments, entwining with the lake's dynamics in ways unforeseen. The narrative of diversion often intersected with the broader regional water management context, creating a nexus where the outcome was not solely determined by the lake's boundaries, but by a complex web of factors beyond.

These previous endeavors, while marked by valiant efforts, have woven a fabric of insights that illuminate the intricate challenges that restoration endeavors must navigate. The Salton Sea's ecosystem, a canvas painted with layers of interdependence, resists swift and straightforward solutions. The lessons from the past beckon towards a nuanced approach, where the threads of ecological balance are delicately rewoven, accounting for the grand tapestry that stretches beyond the lake's shores.

As the pages of history turn, the chapter of restoration continues to unfold, influenced by the trials, tribulations, and triumphs of preceding efforts. The endeavor to rejuvenate the Salton Sea remains an intricate mosaic, enriched by the stories of the past, poised to inform the pathway forward towards a future where vitality is restored, and the echoes of restoration resound through the Salton Sea's narrative once more.

2.4 Atmospheric Water Generation Technology

In the pursuit of innovative solutions to water scarcity, Atmospheric Water Generation (AWG) technology has emerged as a groundbreaking approach. AWG technology harnesses the latent moisture present in the air, converting it into valuable freshwater resources. This technology holds the potential to revolutionize water sourcing strategies, particularly in arid and water-stressed regions such as Southern California. By exploring the application of AWG technology to the restoration of the Salton Sea, a novel avenue for mitigating water scarcity challenges and reducing the dependency on extensive water diversions can be envisioned.

2.4.1 Essence of Atmospheric Water Generation

At its heart, the principle driving Atmospheric Water Generation (AWG) technology is rooted in the natural occurrence of condensation—a phenomenon where the warmth-laden air surrenders its moisture when cooled to a certain degree. This transformative process unfolds through a series of sequential stages, each contributing to the conversion of atmospheric moisture into precious liquid water. The orchestration involves the artful choreography of air intake, cooling, condensation, meticulous filtration, and the reservoir's embrace.

In the symphony of AWG, as air finds its way into the system, a ballet of cooling commences. The delicate balance of temperature prompts the liberated moisture to form glistening droplets, adhering to specially-designed surfaces primed for this aqueous embrace. The resultant liquid gems, imbued with newfound purity, pass through the meticulous filtration choreography, where impurities are ushered away. Finally, the distilled water finds its haven within the reservoir's sanctuary, poised for utilization.

2.4.2 Suitability for Arid Environments

One of the most compelling aspects of AWG technology is its adaptability to arid and semi-arid environments. In regions like Southern California, where traditional water sources are strained and unpredictable, AWG offers a localized and sustainable solution. The technology's effectiveness is not solely dependent on the presence of natural bodies of water or established water infrastructure. Instead, it taps into the ubiquitous resource of atmospheric moisture, making it particularly relevant for areas characterized by water scarcity.

2.4.3 Envisaged Advantages for the Restoration of Salton Sea

Envisioning the incorporation of Atmospheric Water Generation (AWG) technology within the grand tapestry of Salton Sea's restoration imparts a multitude of benefits. The infusion of AWG technology into the restoration symphony ushers in a harmonious transformation. By orchestrating the generation of freshwater from the very fabric of the atmosphere, the reliance on expansive water diversions begins to wane, as if receding tides. This reduction in the scale of diversions orchestrates a delicate ballet, a choreography that reverberates through the ecosystem, refraining from disrupting the cadence of natural water flows.

The allure of AWG's artistry extends beyond mere water generation; it paints a portrait of sustainability and adaptability. With AWG as a steadfast companion, a decentralized source of water unfurls its canvas, adaptable to the ever-changing rhythms of the ecosystem's heartbeat. In this watercolor of restoration, AWG adorns itself with multiple feathers—a reduction in ecological turbulence, a nurturing embrace for the ecosystem, and a lifeline to neighboring communities.

2.4.4 Considerations and Challenges

While AWG technology presents promising opportunities, it is essential to acknowledge potential challenges. Factors such as energy consumption, equipment maintenance, and the scalability of AWG systems require careful evaluation. Additionally, the economic viability and cost-effectiveness of large-scale AWG implementation demand scrutiny.

2.4.5 Synergies with Innovative Infrastructure

Amidst the tapestry of restoration possibilities, the concept of floating solar islands takes center stage as a symphony of innovation and sustainability. Picture a stage where solar panels glide atop the waters of the Salton Sea, harnessing the sun's energy while casting their shadows on the surface below. These floating solar islands, like performers on a global stage, offer a complementary rhythm to the AWG technology's tune.

By integrating floating solar islands with AWG technology, a harmonious duet unfolds. The solar panels, basking in the sun's embrace, generate the energy needed for the cooling and condensation processes of AWG. As the panels serenade the sun's rays, the AWG system orchestrates the production of freshwater, a resource vital to restoring the lake's equilibrium. This synergy transforms the Salton Sea's vast expanse into a canvas of innovation—a canvas where sustainability and restoration interlace their dance.

Yet, as with any production, the backdrop isn't without its complexities. Engineering these solar islands to endure the lake's unique conditions, optimizing their efficiency, and addressing potential environmental impacts all form part of this intricate choreography. The cast includes considerations of spatial arrangements, wildlife interactions, and economic feasibility. Thus, as we envisage these floating solar islands, we step into a world where innovation and restoration join hands, and the stage is set for progress.

2.5 Potential Benefits of Fresh Water Addition

Adding one million liters of fresh water per day to the Salton Sea offers potential benefits such as reducing salinity stress on aquatic organisms, improving water quality, and fostering the recovery of native species. This intervention could create a more favorable environment for both aquatic and avian species, contributing to the restoration of ecosystem services and enhancing the lake's ecological resilience.

By synthesizing existing literature on the Salton Sea's ecosystem challenges, the role of salinity, previous restoration efforts, AWG technology, and the potential benefits of fresh water addition, this chapter sets the foundation for evaluating the feasibility and implications of introducing fresh water as a restoration strategy for the Salton Sea. It highlights the need for a comprehensive approach that considers both the ecological intricacies of the lake and the potential of innovative technologies to address its complex ecological issues.

Chapter 3: Methodology

3.1 Data Collection and Sources

In order to construct a robust and accurate understanding of the current conditions at the Salton Sea, this research undertakes a comprehensive data collection effort from diverse sources. The amalgamation of hydrological and ecological data is integral to developing a holistic picture that informs the subsequent analyses and modeling. Various types of data, ranging from hydrological parameters to ecological indicators, will be sourced from reliable governmental bodies, research institutions, and established environmental monitoring stations.

3.1.1 Hydrological Data Collection

The hydrological data collection process encompasses multiple dimensions of the Salton Sea's water dynamics. Parameters to be collected include:

- **Water Inflow and Outflow Rates:** Governmental agencies responsible for water management, such as the California Department of Water Resources and the Imperial Irrigation District, will provide data on the volume of water flowing into and out of the lake. This data is essential for quantifying the net changes in water levels over time.

- **Evaporation Rates:** Evaporation plays a pivotal role in the water balance of the Salton Sea. Evaporation rates will be obtained from local meteorological stations and environmental monitoring agencies.

- **Historical Water Level Trends:** Records of historical water level fluctuations over time are crucial for understanding the lake's behavior and trends. These records will be sourced from historical data archives and research institutions.

3.1.2 Ecological Data Collection

The ecological component of data collection involves capturing the intricate dynamics of the lake's ecosystem. The following types of ecological data will be collected:

- **Species Composition:** Comprehensive surveys and studies conducted by environmental agencies, research organizations, and academic institutions will provide insights into the composition of aquatic species inhabiting the Salton Sea. These surveys capture data on both native and non-native species.

- **Water Quality Indicators:** Parameters such as nutrient levels, pH, dissolved oxygen, and other water quality indicators will be gathered from environmental monitoring programs and relevant studies. These indicators shed light on the health of the aquatic environment.

- **Habitat Distribution:** Geographical Information Systems (GIS) data will be used to map the distribution of different habitats around the lake. This data is critical for understanding the existing ecological niches and potential impacts of the freshwater addition.

By sourcing data from reputable and authoritative entities, this research aims to ensure the accuracy and reliability of the information used to inform subsequent modeling and analyses. The integration of hydrological and ecological datasets forms the foundation for a comprehensive understanding of the Salton Sea's current state and the potential implications of the proposed restoration strategy.

3.2 Hydrological Modeling

To comprehensively assess the potential impact of introducing one million liters of fresh water per day into the Salton Sea, a hydrological model will be developed. This model will utilize historical data on water inflow rates, outflow rates, evaporation rates, and water level fluctuations to simulate the lake's response to the proposed intervention.

3.2.1 Data Collection and Preparation

Historical data on water inflow rates from surrounding rivers, outflow rates through evaporation and seepage, evaporation rates influenced by meteorological conditions, and water level fluctuations over time will be collected from reputable sources. This data will be carefully curated, cleaned, and standardized to ensure accuracy and reliability in the modeling process.

3.2.2 Development of the Hydrological Model

A dynamic hydrological model will be constructed using mathematical equations and numerical methods. The model will consider the temporal variations in water inflows, outflows, and evaporation rates. Equations describing hydrological processes such as precipitation, inflow from surrounding rivers, evaporation, and outflow will be integrated into the model's framework.

Here's an example of how the mathematical models for the hydrological processes could be represented in the hydrological model:

1. *Precipitation and Inflow from Rivers:*

The rate of water entering the lake through precipitation and inflow from rivers can be modeled as:

```
Inflow_rate = Precipitation_rate + River_inflow_rate
```

2. *Evaporation:*

Evaporation is a key factor affecting the water balance of the lake. It can be modeled using methods like the Penman-Monteith equation:

```
Evaporation_rate = K * (T - T_min) * (RH * es - ea) /
(T_max - T_min)
```

Where:
- `K` is the crop coefficient
- `T` is the average daily temperature
- `T_min` and `T_max` are the minimum and maximum daily temperatures
- `RH` is the relative humidity
- `es` is the saturation vapor pressure
- `ea` is the actual vapor pressure

3. *Outflow and Seepage:*

The outflow from the lake due to evaporation and seepage can be modeled as:

```
Outflow_rate = Evaporation_rate + Seepage_rate
```

Seepage can be modeled using Darcy's law:

```
Seepage_rate = K * A * (H_lake - H_groundwater) / L
```

Where:
- `K` is the hydraulic conductivity of the soil
- `A` is the cross-sectional area
- `H_lake` is the water level in the lake
- `H_groundwater` is the groundwater level
- `L` is the distance between the lake and the groundwater table

4. *Change in Water Level:*

The change in water level over time can be modeled using the difference between inflow and outflow rates:

```
dH/dt = Inflow_rate - Outflow_rate
```

These are simplified mathematical representations of the hydrological processes involved in the model. The actual model would likely involve more complex equations and parameters to accurately capture the interactions within the Salton Sea's water balance. The model would be validated and calibrated using historical data to ensure its accuracy in simulating real-world conditions.

Here's how you can implement simulation and validation using the simplified hydrological model example provided earlier:

```python
import math

# Constants and initial conditions (same as before)
# ...

# Simulation settings
days = 365  # Number of simulation days
freshwater_addition = 1000000  # Freshwater addition
per day (liters)

# Initialize variables
H_lake_simulated = [H_lake_initial] * days
H_groundwater_simulated = [H_groundwater_initial] *
days

# Simulate hydrological processes
for day in range(1, days):
    # Evaporation calculation using Penman-Monteith
equation
    # ...

    # Seepage calculation using Darcy's law
    # ...

    # Inflow rate (assuming constant)
    # ...
```

```python
    # Outflow rate
    # ...

    # Change in water level
    dH = Inflow_rate - Outflow_rate +
(freshwater_addition / A)

    # Update water levels
    H_lake_simulated[day] = H_lake_simulated[day - 1]
+ dH
    H_groundwater_simulated[day] =
H_groundwater_simulated[day - 1] + (Seepage_rate * L)
/ (K_soil * A)

# Validation using historical data
# (Assuming historical data for water levels is
available)
H_lake_historical = [5, 5.2, 5.4, ...]  # Example
historical water levels (m)

# Calculate validation error (root mean squared error)
validation_error =
math.sqrt(sum((H_lake_simulated[day] -
H_lake_historical[day])**2 for day in range(days)) /
days)

# Results
print("Final simulated water level in lake:",
H_lake_simulated[-1])
print("Validation error:", validation_error)
```

In this code, we simulate the hydrological processes for a specified number of days. After the simulation, we compare the simulated water levels with historical data (assumed to be available) to calculate a validation error, which indicates how well the model's predictions match the actual historical observations.

Please note that this is a simplified validation approach. In reality, validation involves more sophisticated statistical methods and may require additional data processing and analysis. Additionally, incorporating historical data for other variables such as inflow rates, evaporation rates, and seepage rates would further enhance the model's accuracy.

3.2.3 Simulation and Validation

The hydrological model will be validated by comparing its outputs with historical observations of water levels, salinity levels, and other relevant hydrological parameters. The model will be adjusted and calibrated to accurately reproduce past conditions, ensuring its reliability in simulating real-world scenarios.

Here's how you can implement simulation and validation using the simplified hydrological model example I provided earlier:

```python
import math

# Constants and initial conditions (same as before)
# ...

# Simulation settings
days = 365  # Number of simulation days
freshwater_addition = 1000000  # Freshwater addition
per day (liters)

# Initialize variables
H_lake_simulated = [H_lake_initial] * days
H_groundwater_simulated = [H_groundwater_initial] *
days

# Simulate hydrological processes
for day in range(1, days):
    # Evaporation calculation using Penman-Monteith
equation
    # ...

    # Seepage calculation using Darcy's law
```

```python
    # ...

    # Inflow rate (assuming constant)
    # ...

    # Outflow rate
    # ...

    # Change in water level
    dH = Inflow_rate - Outflow_rate +
(freshwater_addition / A)

    # Update water levels
    H_lake_simulated[day] = H_lake_simulated[day - 1]
+ dH
    H_groundwater_simulated[day] =
H_groundwater_simulated[day - 1] + (Seepage_rate * L)
/ (K_soil * A)

# Validation using historical data
# (Assuming historical data for water levels is
available)
H_lake_historical = [5, 5.2, 5.4, ...]  # Example
historical water levels (m)

# Calculate validation error (root mean squared error)
validation_error =
math.sqrt(sum((H_lake_simulated[day] -
H_lake_historical[day])**2 for day in range(days)) /
days)

# Results
```

```
print("Final simulated water level in lake:",
H_lake_simulated[-1])
print("Validation error:", validation_error)
```

In this code, we simulate the hydrological processes for a specified number of days. After the simulation, we compare the simulated water levels with historical data (assumed to be available) to calculate a validation error, which indicates how well the model's predictions match the actual historical observations.

Please note that this is a simplified validation approach. In reality, validation involves more sophisticated statistical methods and may require additional data processing and analysis. Additionally, incorporating historical data for other variables such as inflow rates, evaporation rates, and seepage rates would further enhance the model's accuracy.

3.2.4 Introducing Fresh Water Addition

The validated hydrological model will be extended to incorporate the proposed introduction of one million liters of fresh water per day into the Salton Sea. The model will simulate how this additional water input interacts with existing hydrological processes, including inflows, outflows, and evaporation rates.

Here's how you can incorporate the proposed fresh water addition into the hydrological model:

```python
import math

# Constants, initial conditions, and simulation
settings (same as before)
# ...

# Initialize variables
H_lake_simulated = [H_lake_initial] * days
H_groundwater_simulated = [H_groundwater_initial] *
days

# Simulate hydrological processes with fresh water
addition
for day in range(1, days):
    # Evaporation calculation using Penman-Monteith
equation
    # ...

    # Seepage calculation using Darcy's law
    # ...
```

```
    # Inflow rate (assuming constant)
    # ...

    # Outflow rate
    # ...

    # Change in water level with fresh water addition
    dH = Inflow_rate - Outflow_rate +
(freshwater_addition / A) - Evaporation_rate -
Seepage_rate

    # Update water levels
    H_lake_simulated[day] = H_lake_simulated[day - 1]
+ dH
    H_groundwater_simulated[day] =
H_groundwater_simulated[day - 1] + (Seepage_rate * L)
/ (K_soil * A)

# Validation using historical data (same as before)
# ...

# Results
print("Final simulated water level in lake:",
H_lake_simulated[-1])
print("Validation error:", validation_error)
```

In this section of the code, the fresh water addition is incorporated into the simulation by subtracting the effects of evaporation and seepage from the net change in water level. This reflects the fact that the added fresh water contributes to increasing the water level, counteracting the effects of evaporation and seepage.

Remember that this example is still a simplified representation, and a more comprehensive model would consider a wider range of factors and incorporate additional data sources for a more accurate representation of the system.

Here are the complete formulas for each step in the hydrological model, incorporating the proposed fresh water addition:

```python
import math

# Constants
K = 0.5  # Crop coefficient
T_min = 20  # Minimum temperature (°C)
T_max = 30  # Maximum temperature (°C)
RH = 0.7  # Relative humidity
A = 1000  # Cross-sectional area of lake (m²)
L = 10  # Distance between lake and groundwater table
(m)
K_soil = 0.001  # Hydraulic conductivity of soil (m/s)
```

```python
# Initial conditions
H_lake_initial = 5  # Initial water level (m)
H_groundwater_initial = 10  # Initial groundwater
level (m)

# Simulation settings
days = 365  # Number of simulation days
freshwater_addition = 1000000  # Freshwater addition
per day (liters)

# Initialize variables
H_lake_simulated = [H_lake_initial] * days
H_groundwater_simulated = [H_groundwater_initial] *
days

# Simulate hydrological processes with fresh water
addition
for day in range(1, days):
    # Evaporation calculation using Penman-Monteith
equation
    T_avg = (T_min + T_max) / 2
    es = 0.6108 * math.exp((17.27 * T_avg) / (T_avg +
237.3))
    ea = RH * es
    Evaporation_rate = K * (T_avg - T_min) * (RH * es
- ea) / (T_max - T_min)

    # Seepage calculation using Darcy's law
    Seepage_rate = (K_soil * A * (H_lake_simulated[day
- 1] - H_groundwater_simulated[day - 1])) / L
```

```python
    # Inflow rate (assuming constant)
    Inflow_rate = 5000  # Example inflow rate from
rivers (liters/day)

    # Outflow rate
    Outflow_rate = Evaporation_rate + Seepage_rate

    # Change in water level with fresh water addition
    dH = Inflow_rate - Outflow_rate +
(freshwater_addition / A) - Evaporation_rate -
Seepage_rate

    # Update water levels
    H_lake_simulated[day] = H_lake_simulated[day - 1]
+ dH
    H_groundwater_simulated[day] =
H_groundwater_simulated[day - 1] + (Seepage_rate * L)
/ (K_soil * A)

# Validation using historical data
# Calculate validation error (root mean squared error)
validation_error =
math.sqrt(sum((H_lake_simulated[day] -
H_lake_historical[day])**2 for day in range(days)) /
days)

# Results
print("Final simulated water level in lake:",
H_lake_simulated[-1])
print("Validation error:", validation_error)
```

Please note that these formulas are based on the simplified model assumptions and may not capture all the complexities of real-world hydrological systems. Also, be sure to adapt the formulas and code to suit your specific dataset and research objectives.

Here's an example of how you can run multiple simulation scenarios with varying levels of fresh water addition and climatic conditions using the hydrological model:

```python
import math

# Constants, initial conditions, and simulation
settings (same as before)
# ...

# List of different freshwater addition rates to
simulate (liters/day)
freshwater_rates = [0, 500000, 1000000, 1500000]

# Initialize variables for each scenario
scenario_results = []

# Simulate hydrological processes for each scenario
for freshwater_addition in freshwater_rates:
    # Initialize variables for the current scenario
    H_lake_simulated = [H_lake_initial] * days
```

```python
    H_groundwater_simulated = [H groundwater_initial]
* days

    # Simulate hydrological processes with fresh water
addition
    for day in range(1, days):
        # Evaporation calculation using
Penman-Monteith equation
        # ...

        # Seepage calculation using Darcy's law
        # ...

        # Inflow rate (assuming constant)
        # ...

        # Outflow rate
        # ...

        # Change in water level with fresh water
addition
        # ...

        # Update water levels
        # ...

    # Calculate validation error for the scenario
(same as before)
    validation_error =
math.sqrt(sum((H_lake_simulated[day] -
H_lake_historical[day])**2 for day in range(days)) /
days)
```

```
    # Store scenario results
    scenario_results.append({
        "Freshwater Addition": freshwater_addition,
        "Validation Error": validation_error,
        "Simulated Water Levels": H_lake_simulated
    })

# Display scenario results
for scenario in scenario_results:
    print("Scenario: Freshwater Addition =",
scenario["Freshwater Addition"])
    print("Validation Error:", scenario["Validation
Error"])
    print("Final simulated water level in lake:",
scenario["Simulated Water Levels"][-1])

print("-------------------------------------------------")
```

In this code, we define a list of different freshwater addition rates to simulate (`freshwater_rates`). For each rate, we run the hydrological model simulation as before. The results of each scenario, including the validation error and final simulated water levels, are stored in the `scenario_results` list. The code then displays the results for each scenario.

This approach allows you to examine how different levels of freshwater addition impact the hydrological system under varying conditions. Keep in mind that this is a simplified example, and real-world scenarios would involve more complex variations and factors to consider.

3.2.5 Scenario Analysis and Interpretation

Multiple simulation scenarios will be run using the hydrological model, reflecting different levels of fresh water addition and varying climatic conditions. The model's outputs will provide predictions of how the lake's water levels, salinity, and overall water balance might evolve over time under different intervention scenarios. The results will be analyzed to understand potential outcomes, challenges, and implications.

By utilizing historical data on water inflow rates, outflow rates, evaporation rates, and water level fluctuations, this hydrological modeling approach aims to offer insights into how the proposed fresh water addition could impact the

Salton Sea's water balance. The model's predictions will serve as a foundation for subsequent ecological and socioeconomic assessments, supporting informed decision-making for the restoration strategy.

3.3 Ecological Impact Assessment

The infusion of substantial quantities of freshwater into the Salton Sea ecosystem presents a pivotal juncture with far-reaching ecological ramifications. This section delineates the methodology employed to rigorously assess the potential ecological repercussions across dimensions encompassing biodiversity, water quality, and the intricate equilibrium of the ecosystem.

3.3.1 Biodiversity Assessment

Biodiversity is a fundamental indicator of ecosystem health and resilience. The ecological impact assessment focuses on comprehending how the introduction of freshwater might influence the species composition and abundance within the Salton Sea. The assessment will involve:

- **Species Diversity:** Evaluation of changes in species richness and evenness to gauge alterations in the diversity of aquatic life forms.

- **Population Dynamics:** Examination of the population trends of native and non-native species, shedding light on potential shifts in dominance and rarity.

- **Habitat Suitability**: Analysis of habitat preferences and potential shifts in habitat suitability for various species in response to altered salinity levels.

3.3.2 Water Quality Analysis

Water quality is intricately linked to the health of an aquatic ecosystem. This analysis investigates how the influx of freshwater might influence key water quality parameters. The assessment will encompass:

- **Nutrient Levels:** Examination of the impact on nutrient concentrations, considering implications for primary productivity and potential eutrophication.

- **pH and Dissolved Oxygen:** Evaluation of changes in pH and dissolved oxygen levels, crucial for aquatic organisms' physiological processes.

- **Pollutant Dilution:** Assessment of the potential dilution of pollutants and contaminants, elucidating the improvement in water quality and potential mitigation of harmful substances.

3.3.3 Ecosystem Equilibrium Evaluation

The ecosystem's equilibrium is a delicate balance influenced by various interrelated factors. This evaluation aims to discern how the introduction of freshwater might alter trophic interactions, food web dynamics, and overall ecosystem stability. The assessment involves:

- **Trophic Cascades:** Examination of how changes in species abundance might propagate through trophic levels, potentially leading to cascading effects.

- **Food Web Resilience:** Analysis of food web structures and resilience, considering how shifts in species composition might impact energy flows and interactions.

- **Invasive Species Dynamics:** Exploration of the potential for altered conditions to facilitate the establishment and spread of invasive species, potentially disrupting existing ecological relationships.

By embracing a multidimensional ecological impact assessment, this research endeavors to illuminate the potential consequences of the freshwater addition. The insights garnered from this analysis play a pivotal role in steering the restoration initiative towards a balanced and informed ecological revitalization of the Salton Sea.

3.3.1 Data Collection and Baseline Assessment

To conduct an ecological impact assessment, baseline data on the current state of the Salton Sea ecosystem must be collected. This includes information on species composition, water quality parameters, nutrient levels, and habitat conditions. Historical data and expert surveys will aid in establishing a comprehensive understanding of the ecosystem's current status.

3.3.2 Habitat Modeling

Using geographic information systems (GIS) and ecological modeling techniques, habitat models will be created to simulate the potential changes in habitat availability and suitability as a result of the freshwater addition. These models will take into account factors such as water depth, temperature, salinity, and substrate composition. The impact on aquatic vegetation, fish spawning grounds, and bird habitats will be assessed.

Habitat modeling is a crucial aspect of the ecological impact assessment, aimed at understanding how the introduction of freshwater will influence the availability and suitability of habitats within the Salton Sea ecosystem. This section outlines the methodology for creating habitat models and simulating potential changes in habitat dynamics.

Data Collection and Preparation

- Geographic Information Systems (GIS) data will be utilized to gather relevant spatial data, including bathymetric maps, substrate composition, water depth, and current habitat distribution.

- Remote sensing data, satellite imagery, and aerial photographs will be collected to aid in identifying and characterizing different habitats within the ecosystem.

Habitat Classification

- Habitats will be classified based on their physical characteristics, such as water depth, substrate type, and vegetation cover.

- Different habitats, such as shallow marshes, open water areas, and submerged vegetation zones, will be delineated using GIS tools.

Ecological Modeling

- Ecological models will be developed to simulate the response of different habitats to changes in water level, salinity, and other environmental variables resulting from the freshwater addition.

- Habitat suitability models will be created to predict the preferences of various species for specific habitat types. These models will consider factors like water temperature, nutrient availability, and substrate characteristics.

Scenario Development

- Various scenarios representing different levels of freshwater addition and potential changes in water quality will be developed.

- Scenarios will account for variations in freshwater inflow rates, taking into consideration both the gradual increase and sudden changes in the water level.

Hydrodynamic Modeling

- Hydrodynamic models will simulate water circulation patterns within the lake, considering the influence of wind, currents, and water temperature.

- These models will provide insights into how freshwater addition will affect water movement, potentially influencing habitat distribution and mixing.

Model Validation and Calibration

- The habitat models will be validated using existing data on habitat distribution, species composition, and environmental variables.

- The models will be refined and calibrated to ensure they accurately reflect the observed habitat dynamics.

Simulation and Analysis

- Using the developed habitat models and validated scenarios, simulations will be conducted to predict changes in habitat availability and suitability over time.

- The simulations will provide insights into how the introduction of freshwater might alter the spatial distribution of habitats and potentially impact species that rely on specific habitats for feeding, breeding, and shelter.

Integration with Overall Impact Assessment

- The results from the habitat modeling will be integrated with other aspects of the ecological impact assessment, such as biodiversity assessment and water quality monitoring.

- This integrated approach will offer a comprehensive understanding of how the ecosystem's response to freshwater addition will influence the overall ecological balance of the Salton Sea.

By employing geographic information systems and ecological modeling techniques, the habitat modeling process will contribute valuable insights into the potential changes in habitat availability and suitability as a consequence of the proposed freshwater addition. These insights will aid in understanding the potential impacts on various species and guide management decisions to ensure the long-term health and sustainability of the Salton Sea ecosystem.

3.3.3 Biodiversity Assessment

The intricate tapestry of the Salton Sea's biodiversity is a delicate dance of species, each playing a unique role in the lake's ecosystem. To unravel the impact of introducing fresh water, a symphony of biodiversity assessment unfolds—a performance that requires careful monitoring, data collection, and interpretation.

At the heart of this assessment are surveys and expeditions, where scientists and researchers traverse the lake's waters and shores, armed with keen observation and meticulous record-keeping. Their mission: to track the footsteps of both native and non-native species. The aquatic stage hosts a diverse cast of characters, from fish that navigate the depths to birds that grace the skies. Through their presence, or absence, patterns emerge—a shift in distribution, an alteration in abundance, a change in community dynamics.

Fish populations, some native, some introduced, take center stage. These underwater performers interact in a dance of predation, competition, and symbiosis. The surveys track their movements, their numbers, their behaviors—a choreography that tells the story of their response to the freshwater infusion. As the curtain rises, avian actors join the narrative. Birds, whose wings grace both the Salton Sea and distant lands, form a bridge between ecosystems. Monitoring their presence, their nesting habits, their migratory journeys, offers a glimpse into the lake's health and the broader ecological connections.

The data collected in these assessments form the notes of a musical score—an orchestration of information that will be analyzed and interpreted. Are native species finding renewed vitality in the altered conditions? Are non-natives asserting their dominance? How do these changes resonate through the food web, the intricate interplay of who eats whom?

Through biodiversity assessment, the symphony of the Salton Sea's ecosystem becomes clearer—a melody of interactions, a rhythm of adaptation. As researchers gather insights, they contribute to the evolving narrative of restoration, each note a step toward understanding the dance that unfolds when human intention meets the forces of nature.

3.3.4 Water Quality Monitoring

Within the intricate tapestry of the Salton Sea's ecosystem, water quality acts as both the canvas and the brushstroke—a palette of parameters that shape the lives of its inhabitants. To decipher the impact of introducing freshwater, a vigilant eye turns to the lake's chemistry—an observation that delves into the very essence of its waters.

In this monitoring endeavor, scientists become stewards of measurement devices, instruments that peer beneath the surface to reveal the stories told by elements and compounds. Salinity, that silent conductor of aquatic life, is measured in precise quantities—a barometer of the lake's character. pH, the delicate balance of acidity and alkalinity, paints a portrait of the water's temperament. Dissolved oxygen, that giver of life, is quantified, ensuring a hospitable haven for aquatic organisms.

The parameters extend beyond the basic notes of the composition. Nutrient concentrations form a subplot—a tale of excess and scarcity, of the delicate balance between nourishment and overabundance. Heavy metals, those silent actors lurking in the depths, emerge from the shadows through careful analysis—a reminder of the interconnectedness between human activities and the lake's health.

As days turn into seasons, the monitoring continues—a watchful eye on the water's changing hues. Patterns emerge, revealing the impacts of freshwater infusion on the lake's chemistry. Is the salinity melody shifting? Is the pH harmony finding a new balance? Are the nutrients orchestrating a different dance?

Through water quality monitoring, the composition of the Salton Sea's waters becomes a story told in measurements—a narrative of transformation, adaptation, and response. The data collected forms the brushstrokes of understanding, creating a portrait of the lake's evolving relationship with the introduction of fresh water. And as the data unfolds, it contributes to the symphony of knowledge that guides the restoration, ensuring that the melody of the lake's chemistry harmonizes with the notes of ecological rejuvenation.

3.3.5 Predictive Modeling

Utilizing the collected data and advanced ecological modeling techniques, predictive models will be developed to simulate how the introduction of freshwater might affect the ecosystem over time. These models will help forecast potential changes in species composition, population dynamics, and habitat suitability under varying scenarios.

3.3.6 Sensitivity Analysis

Sensitivity analysis will involve subjecting the ecological models to different freshwater addition rates, climatic conditions, and potential scenarios. This analysis aims to identify critical thresholds beyond which ecological impacts become significant, helping establish precautionary measures and management strategies.

3.3.7 Stakeholder Consultation

Throughout the ecological impact assessment, engagement with stakeholders, including environmental agencies, local communities, and conservation organizations, will be crucial. Stakeholder input will provide valuable insights, aid in identifying potential concerns, and ensure a collaborative approach to decision-making.

By conducting a comprehensive ecological impact assessment, the study will provide valuable insights into the potential consequences of the proposed freshwater addition. These insights will guide policymakers, conservationists, and stakeholders in making informed decisions that prioritize the preservation and restoration of the Salton Sea's unique and fragile ecosystem.

3.4 Economic and Societal Considerations

Beyond the ecological dimensions, the research embarks on a comprehensive exploration of the economic and societal ramifications arising from the introduction of freshwater to the Salton Sea. This multifaceted analysis delves into the potential economic advantages for local communities, the cascading effects of improved water quality on broader economies, and the imperative role of public sentiment and stakeholder engagement in shaping the restoration endeavor.

3.4.1 Economic Benefits for Local Communities

The Salton Sea's rejuvenation has the potential to catalyze economic growth at the local level. This facet of the analysis evaluates how the influx of freshwater might yield economic dividends, including but not limited to:

- **Tourism and Recreational Opportunities:** The restoration of a healthier aquatic ecosystem could attract tourists and recreational enthusiasts, generating revenue through increased visitation, accommodations, and related activities.

- **Local Businesses and Employment:** Enhanced tourism and recreational activities may spur the growth of local businesses, creating job opportunities and bolstering economic vitality.

- **Property Values:** The revitalization of the Salton Sea could enhance property values in nearby areas, benefiting homeowners and local property markets.

3.4.2 Broader Economic Implications

The research extends its gaze beyond local communities to scrutinize the broader economic implications of an ecologically restored Salton Sea. Key aspects of this analysis include:

- **Water Quality-Driven Economic Boost:** Improved water quality has the potential to alleviate costs associated with healthcare and water treatment due to reduced pollution, ultimately contributing to economic savings.

- **Real Estate and Investments:** A rehabilitated Salton Sea might attract new investments, potentially fostering a conducive environment for economic growth across sectors.

3.4.3 Public Perception and Stakeholder Engagement

The success of restoration endeavors pivots on the nexus of public acceptance and stakeholder engagement. This facet of the research underscores:

- **Public Perception:** The analysis delves into the perception of local residents, environmental organizations, governmental bodies, and the broader public concerning the restoration initiative. Identifying concerns and sentiments is paramount for informed decision-making.

- **Stakeholder Engagement:** This aspect explores strategies for engaging stakeholders, nurturing transparent dialogues, and fostering shared ownership. Effective engagement fosters a sense of collaboration, leading to a more cohesive and successful restoration process.

By venturing into the terrain of economic and societal considerations, this research comprehensively embraces the intricate web of factors that influence the feasibility, sustainability, and societal acceptance of the proposed freshwater addition to the Salton Sea.

By adopting a methodological approach that integrates data collection, hydrological modeling, ecological impact assessment, and considerations of economic and societal factors, this chapter outlines the systematic approach to evaluating the effects of introducing fresh water into the Salton Sea. The methodology aims to provide a comprehensive understanding of the potential outcomes and implications of this restoration strategy, allowing for informed decision-making and sustainable planning

Chapter 4: Hydrological Assessment

4.1 Current Water Balance and Salinity Levels

This section delves into a comprehensive analysis of the present-day water balance and salinity levels of the Salton Sea. By dissecting the intricate interplay of inflows, outflows, and salinity dynamics, this examination lays the foundation for understanding the lake's current state, a critical precursor to evaluating the potential impact of the proposed freshwater addition.

4.1.1 Inflows and Outflows

The analysis of water balance revolves around deciphering the intricate web of inflows and outflows that govern the Salton Sea's hydrological equilibrium. Key considerations encompass:

- **Agricultural Runoff:** Water originating from agricultural activities, often laden with nutrients and pollutants, constitutes a substantial inflow source. Comprehensive data on the volume and composition of agricultural runoff will be collated from agricultural agencies and monitoring stations.

- **Local Rivers and Streams:** Contributions from local rivers and streams contribute to the lake's water influx. Precise quantification of these inflows, sourced from hydrological databases, enhances the accuracy of the water balance assessment.

- **Evaporation and Seepage:** The study delves into the magnitude of water loss due to evaporation, which is particularly pronounced in arid environments. Additionally, seepage into the groundwater system forms a significant component of the outflows.

4.1.2 Salinity Levels and Water Quality

The section also delves into the salinity levels and water quality parameters characterizing the Salton Sea. This entails:

- **Salinity Trends**: Utilizing historical data, the assessment traces the temporal evolution of salinity levels, capturing fluctuations and trends that have occurred over time.

- **Water Quality Parameters:** Beyond salinity, other water quality parameters, such as nutrient levels, pH, and dissolved oxygen, will be explored to comprehend the ecological conditions within the lake.

By amalgamating data from diverse sources, this analysis endeavors to provide a comprehensive overview of the Salton Sea's water balance and salinity dynamics. The insights garnered serve as a benchmark against which the potential implications of introducing freshwater can be juxtaposed, thereby guiding informed decision-making in the subsequent phases of the restoration endeavor.

4.2 Impact of Fresh Water Addition on Water Levels

Drawing upon the hydrological model meticulously constructed in Chapter 3, this section embarks on a simulation journey to unravel the intricate tapestry of how the introduction of one million liters of fresh water daily might sculpt the water levels of the Salton Sea. With a keen focus on accommodating the nuances of seasonal variability in inflow and evaporation rates, this analysis endeavors to foretell the cascading ramifications that the infusion of freshwater could usher in terms of altered water depth and overarching volume.

4.2.1 Hydrological Modeling and Simulation

The hydrological model, birthed from the synergy of empirical data and mathematical representation, takes center stage. This sophisticated model, forged from the data collated in Chapter 3, captures the dynamic interplay of inflow and outflow, translating it into a predictive framework.

4.2.2 Fresh Water Addition Scenarios

A spectrum of scenarios takes shape, each weaving a unique narrative. The simulation encompasses:

- **Daily Fresh Water Addition:** Employing the one million liters of daily fresh water as a fulcrum, the model quantifies how this incremental input could influence the lake's hydrological equilibrium.

- **Seasonal Variability:** The analysis contemplates the ebb and flow of seasons, acknowledging their distinct inflow and evaporation patterns. This dynamic perspective encapsulates the cyclical shifts that characterize the lake's hydrodynamics.

4.2.3 Projection of Water Depth and Volume Changes

The simulation's essence crystallizes into a projection of water depth and volume alterations. It envisages how the added freshwater interacts with the lake's existing hydrological dynamics, tracing its ripple effect on water levels over time.

By harnessing the predictive power of the hydrological model, this exploration unveils a tapestry of possible futures. The insights garnered enable stakeholders to peer into the horizon and gauge the tangible consequences of introducing freshwater into the Salton Sea. This pivotal understanding sets the stage for informed decision-making, paving the way for the next chapter in the restoration narrative.

4.3 Modeling Evaporation Rates

In the intricate dance of the Salton Sea's hydrological equilibrium, evaporation emerges as a pivotal player, wielding the power to shape water balance dynamics. This section steps onto the stage to demystify the art of evaporation modeling, as well as to explore how the infusion of fresh water might choreograph changes in this phenomenon. By juxtaposing historical climate data against the canvas of freshwater addition, this analysis seeks to illuminate the potential shifts in evaporation dynamics and their reverberations across the lake's water loss landscape.

4.3.1 Harnessing Historical Climate Data

Historical climate data takes center stage as a foundational cornerstone. Temperature, humidity, wind speed, and other meteorological variables, harvested from archives and monitoring stations, serve as the bedrock for the ensuing evaporation modeling.

4.3.2 Evaporation Modeling

Drawing from meteorological data, evaporation modeling finds its form. Mathematical equations, calibrated to capture the intricate interplay of climatic variables, unveil the rate at which water metamorphoses into vapor. These models stand as bridges, spanning the chasm between weather dynamics and water transformation.

4.3.3 Pre- and Post-Intervention Comparison

The crux of the analysis lies in the comparison between the pre- and post-intervention evaporation rates. This juxtaposition unravels the narrative of change, tracing how the introduction of fresh water might orchestrate shifts in the evaporation symphony.

4.3.4 Estimating Water Loss Impact

The estimation of the potential impact on water loss emerges as a cornerstone of this analysis. By discerning how the altered evaporation dynamics under the aegis of freshwater addition affect water loss, stakeholders gain a clearer understanding of the implications for the lake's water balance.

As this section weaves the tale of evaporation modeling, it sets the stage for a deeper comprehension of how freshwater infusion might tweak the choreography of water loss. Armed with insights into evaporation's metamorphosis, the research ascends to the precipice of Chapter 4.4, poised to unravel the dynamic interactions that underlie the restoration narrative.

4.4 Assessing Changes in Inflow and Outflow Patterns

In the delicate choreography of the Salton Sea's hydrological ballet, the introduction of fresh water injects new dynamics, potentially reshaping the interplay between inflow and outflow. This section steps into the spotlight, poised to decipher how shifts in water balance, born from the embrace of freshwater, might orchestrate transformations in the lake's hydrological exchanges. By unraveling the intricate threads that link altered water levels, salinity, and the surrounding web of rivers, streams, and agricultural runoff, this analysis sets out to predict the ensuing changes, ultimately casting light on the broader ripples that the intervention might cascade across the local watershed.

4.4.1 Dynamic Inflow and Outflow Nexus

The analysis takes a holistic stance, casting its gaze upon the intricate nexus of inflow and outflow. Rivers, streams, and agricultural runoff, each a distinct character in the hydrological narrative, interact dynamically with the lake. Their interaction responds to the lake's water levels and salinity, shaping the tapestry of inflow and outflow patterns.

4.4.2 Influence of Altered Water Conditions

As the script unfolds, the focus narrows onto the potential influence of altered water conditions. Changes in water levels and salinity, wrought by the introduction of fresh water, reverberate across the local hydrological landscape, potentially reshaping the interactions between the lake and its surrounding waterways.

4.4.3 Predicting Hydrological Interactions

Guided by empirical data and the hydrological model, predictions come to life. The analysis envisages how changes in water dynamics might translate into tangible shifts in the interactions between the Salton Sea and its neighboring rivers, streams, and agricultural inflows.

4.4.4 Insights into Broader Effects

As the chapter reaches its crescendo, the insights garnered into altered inflow and outflow patterns illuminate the broader effects of the intervention on the local watershed. This understanding resonates as a compass, guiding stakeholders toward comprehensive decision-making that transcends the lake's shores and resonates across the interconnected hydrological fabric.

In this unfolding narrative, the story of altered hydrological interactions takes its rightful place, serving as a vital thread that binds the restoration endeavor to the heartbeat of the surrounding ecosystem.

Chapter 5: Ecological Impact Analysis

5.1 Effects on Aquatic Species and Habitats

In the intricate web of the Salton Sea's ecosystem, the introduction of fresh water emerges as a catalyst of change, potentially painting new strokes on the canvas of aquatic life and habitats. This section takes a plunge into the depths, delving into the realm of possibilities where the infusion of fresh water might script narratives of transformation for the lake's aquatic denizens and their intricate abodes. Armed with ecological models and the reservoir of available data, this chapter embarks on a voyage of exploration, tracing the pathways by which changes in salinity and water quality may influence the distribution, behavior, and survival of the diverse cast of fish, invertebrates, and plant species that grace the aquatic stage. In this symphony of analysis, special attention is granted to those species that bear the mantle of sensitivity to salinity fluctuations.

5.1.1 Ecological Models: Unveiling Insights

Ecological models, akin to lanterns illuminating the dark recesses, guide the way. These intricate frameworks, woven from empirical data and mathematical artistry, unveil the potential effects of fresh water addition on the aquatic tapestry. They spotlight the interplay between salinity, water quality, and the lives that unfold beneath the surface.

5.1.2 Navigating the Impacts

The exploration sweeps through the diverse cast of aquatic inhabitants, from the majestic fish to the subtle invertebrates and the resilient plant species. Each chapter of this analysis delves into how altered salinity and water quality may play a role in their fate.

5.1.3 Dancing with Sensitivity

Amidst the aquatic choreography, sensitivity emerges as a guiding light. Species that dance on the precipice of salinity tolerance are granted special attention. Their delicate balance teeters, and the analysis seeks to unveil how fresh water's embrace might tip the scales of survival or challenge.

5.1.4 From Distribution to Behavior

From the distribution maps that chart species' domains to the behaviors that orchestrate their existence, the analysis sculpts a narrative. It captures how the influx of fresh water might remix the canvas, potentially influencing migrations, feeding patterns, and the very rhythm of life beneath the water's mirror.

In the symphony of change, the chapter stands as a conductor, wielding ecological models and empirical insights to compose a harmonious narrative. With each revelation, it paints strokes of understanding on the canvas of aquatic species and habitats, setting the stage for a comprehensive understanding of the restoration endeavor.

5.2 Potential for Enhanced Biodiversity

In the midst of the Salton Sea's transformational journey, the curtain rises on the prospect of enhanced biodiversity, a promise of nature's resurgence orchestrated by the hands of fresh water. This section steps onto the stage, a torchbearer illuminating the path toward a richer, more diverse tapestry of life. Amidst the cadence of lowering salinity through the infusion of fresh water, whispers of a potential renaissance beckon—an ecosystem teeming with life, both old and new. With a discerning gaze, this chapter unfurls the potential for an enriched biodiversity, spotlighting the possibility of welcoming back native species that once waltzed on the shores and through the waters.

5.2.1 Lower Salinity's Invitation

Fresh water emerges as the harbinger of change, lowering salinity levels and extending an invitation to forgotten inhabitants. The analysis seeks to unravel how this transformation may foster conditions that kindle the re-emergence of native species, a beacon of hope for heightened biodiversity.

5.2.2 The Return of Natives

Amidst the pages of historical records and ecological memory, native species stand as protagonists in a tale of resilience. This section traces their journeys, investigating how the reduction in salinity might nudge them back into the spotlight. It takes a poignant look at species that once suffered the consequences of high salinity levels.

5.2.3 From Shadows to Center Stage

As the script unfolds, the narrative shifts from shadows to center stage. The potential for a revival takes root in the restoration plan, weaving a tapestry where the likes of fish, crustaceans, and aquatic plants may once again grace the stage with their presence.

5.2.4 Nature's Symphony Reimagined

In this reimagined symphony, species interactions, migrations, and habitat utilization are rewritten. The potential for enhanced biodiversity resonates as a melody, where native species find their harmonious place amidst the changing composition of the ecosystem.

The chapter stands as a harbinger of possibility, a beacon that calls forth the prospect of a more vibrant and diverse ecosystem. With each revelation, it fashions a narrative of resurgence, where the echoes of past inhabitants might once again reverberate through the shores and waters of the Salton Sea.

5.3 Mitigation of Salinity-Induced Stress

High salinity levels can lead to stress and physiological challenges for aquatic organisms. This section will explore how the introduced freshwater might alleviate salinity-induced stress, potentially enhancing the overall health and reproductive success of the lake's ecosystem.

The adverse effects of elevated salinity levels on aquatic organisms within the Salton Sea ecosystem are well-documented. This section delves into the potential of introduced freshwater to mitigate the stress induced by high salinity. By exploring how the influx of freshwater might alleviate physiological challenges, enhance ecosystem health, and promote reproductive success, we aim to illuminate a pathway towards restoring the balance of this unique ecosystem.

Understanding Salinity-Induced Stress

- High salinity levels pose a myriad of challenges to aquatic organisms, affecting osmoregulation, metabolic processes, and reproductive success.

- Increased salinity can lead to reduced oxygen availability, disrupted ion balance, and impaired nutrient absorption, impacting overall health and survival.

Alleviating Osmotic Stress

- Freshwater introduction can dilute the salinity levels of the water body, reducing the osmotic stress experienced by aquatic organisms.

- Lower salinity enhances osmoregulation, allowing organisms to maintain optimal internal ion concentrations and physiological functions.

Enhancing Metabolic Efficiency

- Elevated salinity demands additional energy expenditure for osmoregulation, diverting energy from growth, reproduction, and other vital processes.

- Reduced salinity following freshwater addition can alleviate this energy burden, potentially fostering enhanced metabolic efficiency and growth rates.

Promoting Reproductive Success

- High salinity can hinder reproductive success by impacting egg viability, larval development, and the survival of young organisms.

- Lower salinity levels resulting from the freshwater input might create more favorable conditions for reproduction, potentially leading to increased recruitment and population recovery.

Balancing Biotic Interactions

- Salinity-induced stress can alter predator-prey dynamics and disrupt trophic interactions within the ecosystem.

- Reduced salinity could restore natural feeding behaviors, aiding in the recovery of species that were previously adversely affected.

Consideration of Ecological Dynamics

- The mitigation of salinity-induced stress through freshwater addition is complex and must consider the interplay of various ecological factors.

- Research on the responses of key species to changing salinity conditions will guide the assessment of the overall ecological benefits.

Monitoring and Adaptive Management

- Continuous monitoring of salinity levels, organism responses, and ecosystem dynamics is essential to validate the positive outcomes of freshwater addition.

- Adaptive management strategies will be crucial to fine-tune the freshwater input rates based on observed ecological responses.

By exploring the potential of introduced freshwater to mitigate salinity-induced stress, this section sheds light on a critical aspect of the restoration initiative. The delicate balance between salinity and aquatic life is fundamental to the Salton Sea's ecological resilience. Through thoughtful mitigation strategies, we endeavor to restore the ecosystem's health, foster thriving populations, and ensure the longevity of this remarkable aquatic environment.

5.4 Predicting Changes in Ecosystem Dynamics

By synthesizing ecological models and hydrological predictions, this section will offer insights into how changes in salinity and water quality might alter the dynamics of the Salton Sea's ecosystem. The chapter will assess potential shifts in trophic interactions, food web structures, and the overall resilience of the ecosystem to future changes.

The intricate interplay between ecological and hydrological factors shapes the dynamics of the Salton Sea's ecosystem. This section harnesses the synergy between ecological models and hydrological predictions to unveil the potential consequences of altered salinity and water quality. By illuminating potential shifts in trophic interactions, food web structures, and the ecosystem's resilience, we aim to provide a holistic understanding of the ecosystem's response to restoration efforts.

Synthesizing Ecological Models and Hydrological Predictions

- Ecological models will be integrated with hydrological predictions to create holistic simulations of the ecosystem's response to altered salinity levels.

- These models will consider variables such as nutrient availability, temperature, and habitat changes in response to the introduced freshwater.

Shifting Trophic Interactions

- Changes in salinity can influence the distribution and abundance of primary producers, subsequently impacting herbivores, predators, and decomposers.

- Predicted alterations in trophic interactions might lead to shifts in species composition, with potential consequences for ecosystem stability.

Food Web Reshaping

- Altered salinity can modify the availability of prey items for various organisms, potentially leading to changes in feeding behaviors and food preferences.

- The resulting reshaping of the food web structure might cascade through the ecosystem, influencing predator-prey relationships and energy flow.

Resilience and Adaptation

- Insights from ecological models can inform predictions about the ecosystem's resilience to future changes, including variations in climate and hydrology.

- Understanding how species might adapt to modified salinity conditions will guide assessments of long-term ecosystem health.

Influence on Endangered Species

- The predictions will include an assessment of how altered salinity might impact endangered species that rely on the Salton Sea's unique habitat.

- Endangered species recovery strategies will be informed by the anticipated effects of restoration efforts on their habitat and feeding patterns.

Scenario Testing and Management Strategies

- Different scenarios of freshwater introduction and resultant changes in salinity will be tested using the synthesized models.

- Adaptive management strategies will be developed based on the predictions to address unforeseen consequences and optimize restoration outcomes.

Long-Term Monitoring and Adaptation

- Continuous monitoring of ecological indicators and hydrological variables will be essential to validate the predictions and refine the models over time.

- Adaptive management plans will ensure the restoration initiative remains flexible and responsive to emerging insights and observations.

By unifying ecological models and hydrological predictions, this section aims to illuminate the potential transformations that may ripple through the Salton Sea's intricate ecosystem. Predicting changes in trophic interactions, food web dynamics, and overall ecosystem resilience is crucial for informed decision-making and the design of effective management strategies. As we anticipate the multifaceted outcomes of restoration efforts, this integrated approach promises to guide us towards a harmonious and thriving Salton Sea ecosystem.

Through a comprehensive ecological impact analysis, this chapter aims to provide a nuanced understanding of how introducing fresh water to the Salton Sea could influence its aquatic species, habitats, biodiversity, and overall ecosystem dynamics. By considering potential benefits, challenges, and long-term implications, this analysis will contribute to informed decision-making regarding the restoration strategy's ecological feasibility.

Chapter 6: Environmental and Socioeconomic Implications

6.1 Water Quality Improvement and Pollution Reduction

This section will assess how the introduction of fresh water may lead to improvements in water quality by diluting pollutants and contaminants present in the Salton Sea. The potential reduction in nutrient loading, heavy metals, and other pollutants will be examined, along with the resulting benefits for aquatic life and ecosystem health.

The introduction of freshwater presents a promising avenue for enhancing water quality within the Salton Sea ecosystem. This section delves into the potential improvements that can arise from the dilution of pollutants and contaminants currently affecting the water body. By analyzing the anticipated reduction in nutrient loading, heavy metals, and other pollutants, we aim to shed light on the potential benefits for aquatic life, ecosystem health, and overall environmental integrity.

Understanding Current Water Quality Challenges

- The Salton Sea faces significant water quality challenges, including elevated nutrient concentrations, heavy metals, and pollutants stemming from agricultural runoff and industrial sources.

- These contaminants contribute to eutrophication, algal blooms, and ecological stressors that compromise the well-being of the ecosystem.

Dilution of Nutrient Loading

- The influx of freshwater can lead to a reduction in nutrient concentrations, particularly nitrogen and phosphorus, that contribute to excessive algal growth.

- Lower nutrient levels can mitigate the occurrence of harmful algal blooms, which disrupt the aquatic food web and deplete oxygen levels.

Decrease in Heavy Metals and Contaminants

- Freshwater addition has the potential to dilute heavy metals and contaminants, stemming from agricultural runoff and industrial discharges.

- Reduced concentrations of these substances can alleviate toxic stress on aquatic organisms, leading to improved reproductive success and overall health.

Benefits for Aquatic Life

- Enhanced water quality resulting from pollutant reduction can create more favorable conditions for aquatic organisms, supporting growth, survival, and reproduction.

- Improved water clarity and reduced turbidity can foster the growth of submerged aquatic vegetation, benefiting both habitat availability and water filtration.

Ecosystem Health and Resilience

- Enhanced water quality can contribute to the restoration of natural ecological processes, such as nutrient cycling and energy flow.

- A healthier ecosystem is better equipped to withstand natural stressors, including variations in temperature and water availability.

Consideration of Long-Term Impacts

- The potential benefits of improved water quality must be considered in conjunction with other restoration strategies and the evolving hydrological dynamics of the Salton Sea.

- Long-term monitoring and adaptive management are crucial to track changes and adjust restoration efforts based on observed outcomes.

Social and Economic Benefits

- The restoration of water quality contributes to the overall well-being of local communities and recreational users, enhancing the aesthetics and recreational value of the lake.

- Improving water quality can stimulate economic activities such as tourism, fishing, and water-based recreation.

By assessing the potential for water quality improvement and pollution reduction through freshwater introduction, this section underscores the interconnectedness of ecological health and human well-being. The benefits of enhanced water quality extend beyond the aquatic realm, reaching local communities and broader ecosystems. As we strive to restore the Salton Sea's environmental integrity, the realization of cleaner waters and healthier habitats stands as a testament to the power of restoration efforts.

6.2 Economic Benefits for Local Communities

Fresh water addition could have positive economic implications for nearby communities. This section will explore the potential for increased tourism, recreational activities, and the enhancement of local economies. It will also consider the economic value of a healthier ecosystem in terms of improved property values and reduced healthcare costs associated with environmental degradation.

The introduction of fresh water into the Salton Sea ecosystem holds the potential to bring about a range of positive economic outcomes for the surrounding communities. This section delves into the multifaceted economic benefits that can stem from enhanced water quality, thriving ecosystems, and improved natural landscapes. By examining the potential for increased tourism, recreational activities, enhanced local economies, and the valuation of a healthier environment, we aim to underscore the holistic advantages that restoration efforts can offer.

Economic Stimulus through Tourism

- Improved water quality and ecosystem health can attract tourists seeking recreational opportunities, birdwatching, and water-based activities.

- The allure of a revitalized Salton Sea can bring in revenue from visitors, bolstering local businesses such as hotels, restaurants, and retail establishments.

Recreational Opportunities

- The introduction of fresh water can lead to the re-establishment of aquatic life and the return of fish species, enhancing fishing and boating activities.

- Recreation enthusiasts can explore hiking, camping, and birdwatching in a restored ecosystem, contributing to a diversified tourism portfolio.

Local Economy Enhancement

- Economic growth in nearby communities can result from increased spending on accommodation, dining, and local services by visitors and tourists.

- New job opportunities may emerge in the hospitality, ecotourism, and recreational sectors, benefiting residents and stimulating economic development.

Property Value Appreciation

- A healthier and aesthetically pleasing ecosystem can lead to higher property values in areas adjacent to the Salton Sea.

- Real estate markets may experience increased demand as potential homeowners and investors are attracted to the improved natural surroundings.

Healthcare Cost Reduction

- A rejuvenated ecosystem can contribute to decreased healthcare costs by reducing air and water pollution, which are linked to respiratory and other health issues.

- A healthier environment can alleviate the financial burden on local healthcare systems, promoting community well-being.

Community Pride and Identity

- A revitalized Salton Sea can foster a sense of community pride and identity among local residents, who can take pride in the restoration efforts and their positive impacts.

- A healthier environment can contribute to an improved quality of life for residents, enhancing overall community morale.

Sustainable Economic Growth

- The long-term sustainability of economic benefits hinges on responsible management practices that safeguard the restored ecosystem.

- A balanced approach ensures that economic gains are aligned with the preservation of the environment for future generations.

By examining the potential economic benefits for local communities, this section highlights the interconnected relationship between environmental health and economic prosperity. The restoration of the Salton Sea ecosystem can serve as a catalyst for a vibrant and resilient local economy, underpinned by sustainable tourism, enhanced property values, and reduced healthcare costs. As restoration efforts unfold, the confluence of ecological and economic vitality promises to foster enduring benefits for both the environment and the people who call its surroundings home.

6.3 Social Acceptance and Stakeholder Engagement

The successful implementation of any restoration strategy hinges on social acceptance and stakeholder engagement. This section will investigate public attitudes towards the fresh water addition and the perceptions of stakeholders such as local residents, environmental groups, and government agencies. Strategies for effective communication and collaboration will be explored to ensure a smooth implementation process.

The realization of successful restoration strategies rests upon the bedrock of social acceptance and meaningful engagement with stakeholders. This section delves into the intricate landscape of public attitudes towards the introduction of fresh water and the perceptions of crucial stakeholders, including local residents, environmental organizations, and government agencies. By investigating strategies for effective communication, fostering collaboration, and addressing concerns, we aim to navigate the terrain of social dynamics to ensure the seamless implementation of restoration efforts.

Public Attitudes and Perceptions

- Assess public perceptions and attitudes towards the introduction of fresh water, understanding concerns, aspirations, and expectations.

- Recognize that public support is pivotal for the long-term success of restoration initiatives, and consider feedback as an invaluable asset.

Stakeholder Involvement and Representation

- Engage with a diverse range of stakeholders, including local residents, environmental groups, regulatory bodies, and government agencies.

- Incorporate stakeholder input into decision-making processes to ensure a balanced approach that accounts for various perspectives.

Effective Communication Strategies

- Develop clear and accessible communication materials that convey the benefits of fresh water addition, the rationale behind it, and the anticipated outcomes.

- Tailor communication strategies to resonate with different stakeholder groups, addressing their specific concerns and interests.

Community Consultations

- Organize community consultations, workshops, and town hall meetings to provide platforms for stakeholders to voice their opinions, ask questions, and engage in meaningful discussions.

- These consultations facilitate open dialogues that foster trust and enable stakeholders to actively participate in shaping the restoration initiative.

Collaboration and Co-Creation

- Foster collaborative partnerships that go beyond information-sharing and involve stakeholders in co-creating restoration strategies.

- Collaboration can harness local knowledge, expertise, and resources, leading to more holistic and sustainable outcomes.

Addressing Concerns and Mitigating Conflicts

- Address concerns, skepticism, and conflicts proactively through transparent communication, scientific evidence, and respectful engagement.

- Provide opportunities for stakeholders to express their reservations and work collectively towards solutions.

Long-Term Engagement and Ownership

- Maintain continuous engagement with stakeholders throughout the restoration process, keeping them informed about progress, setbacks, and adaptations.

- Instill a sense of ownership among stakeholders, emphasizing that the restoration's success is intertwined with their involvement.

Building Trust and Resilience

- Transparent decision-making, responsiveness to feedback, and accountability build trust and resilience in the face of challenges.

- When stakeholders are part of the decision-making process, they are more likely to champion the restoration initiative.

By exploring the landscape of social acceptance and stakeholder engagement, this section underscores the pivotal role that community involvement and collaboration play in restoration endeavors. The synergy between public support, effective communication, and stakeholder engagement forms the foundation upon which successful restoration initiatives are built. As we traverse the intricate path of restoration, fostering inclusivity and partnership promises to guide us toward a shared vision of a healthier and thriving Salton Sea ecosystem.

6.4 Long-Term Sustainability and Monitoring

To ensure the sustainability of the restoration effort, long-term monitoring and adaptive management are crucial. This section will discuss strategies for ongoing monitoring of water quality, hydrological changes, and ecological responses. It will also address the need for flexibility in the restoration plan to accommodate unforeseen challenges and evolving ecological dynamics.

The enduring success of any restoration initiative hinges on the commitment to long-term sustainability and the vigilant practice of adaptive management. This section delves into the strategies that underpin the continuous monitoring of water quality, hydrological shifts, and ecological responses within the Salton Sea ecosystem. By exploring the necessity of adaptability and the importance of proactive management in the face of unforeseen challenges, we aim to pave the way for the restoration effort's enduring vitality.

Continuous Monitoring of Water Quality

- Establish a comprehensive water quality monitoring program that tracks parameters such as salinity, nutrient concentrations, and pollutant levels.

- Regular monitoring ensures that changes in water quality are identified promptly, enabling timely intervention if needed.

Hydrological Observations and Modeling

- Implement a robust hydrological monitoring system to track water inflow, outflow, and evaporation rates.

- Coupled with hydrological models, observations inform predictions about water level fluctuations and help assess the impact of restoration strategies.

Ecological Response Assessment

- Monitor the response of key species and ecological indicators to the restoration interventions, tracking population dynamics, habitat shifts, and species interactions.

- Ecological observations provide valuable insights into the success of restoration efforts and guide adaptive management decisions.

Adaptive Management Strategies

- Develop adaptive management plans that outline how restoration strategies will be adjusted based on observed outcomes and changing ecological dynamics.

- Flexibility allows for real-time responses to unexpected challenges and the incorporation of new scientific insights.

Incorporating Lessons Learned

- Regularly review and assess the outcomes of restoration actions, identifying both successes and areas for improvement.

- Incorporate lessons learned from previous projects and apply them to refine the ongoing restoration effort.

Stakeholder Engagement in Monitoring

- Include stakeholders in the monitoring process, keeping them informed about progress, challenges, and the need for adaptations.

- Collaboration with stakeholders fosters shared ownership of the restoration initiative and encourages collective problem-solving.

Communicating Monitoring Results

- Transparently communicate monitoring results and management decisions to stakeholders, fostering trust and understanding.

- Keeping stakeholders informed empowers them to contribute to the ongoing success of the restoration initiative.

Long-Term Commitment

- Recognize that long-term sustainability requires ongoing financial, human, and technological resources.

- Institutionalize monitoring efforts to ensure that data collection and management persist beyond the initial stages of restoration.

By prioritizing long-term sustainability and proactive monitoring, this section underscores the commitment required to uphold the restoration initiative's positive impacts. The dynamic interplay between scientific observation, adaptive management, and stakeholder engagement is central to maintaining a resilient and thriving Salton Sea ecosystem. As the journey of restoration unfolds, the lessons learned and insights gained from ongoing monitoring guide us towards the holistic restoration of this unique and valuable environment.

By examining the potential environmental and socioeconomic implications of introducing fresh water to the Salton Sea, this chapter aims to provide a holistic understanding of the broader consequences of the restoration strategy. By considering the positive outcomes for water quality, economic prosperity, and community engagement, the research will highlight the multi-faceted benefits that can arise from effectively restoring the lake's ecosystem.

Chapter 7: Challenges and Risks

7.1 Potential Negative Consequences

While the introduction of fresh water holds promise for restoring the Salton Sea ecosystem, it is imperative to critically assess potential negative consequences that could emerge from this intervention. This section delves into the nuanced evaluation of unintended ecological disruptions and challenges that may arise. By exploring the possibilities of invasive species colonization, shifts in species composition, altered water dynamics, and increased sediment resuspension, we aim to ensure a comprehensive understanding of the complexities surrounding the restoration effort.

Unintended Ecological Disruptions

- The introduction of fresh water might create opportunities for invasive species to colonize the ecosystem, potentially outcompeting native species and altering existing ecological dynamics.

- Invasive species could negatively impact habitat availability, food web interactions, and the overall ecological balance.

Shifts in Species Composition

- Altered salinity levels may trigger shifts in species composition, favoring certain species over others.

- Such shifts could affect the availability of prey items, leading to cascading effects throughout the food web.

Altered Water Dynamics

- Fresh water addition could result in changes in water circulation patterns, potentially impacting nutrient distribution and oxygen availability.

- Altered water dynamics might influence the distribution of aquatic organisms and their ability to find suitable habitats.

Increased Sediment Resuspension

- Changes in water dynamics could lead to increased sediment resuspension, potentially impacting water clarity and light penetration.

- Sediment resuspension might affect the growth of submerged aquatic vegetation and alter the habitat structure.

Potential for Eutrophication

- While freshwater addition may mitigate eutrophication to some extent, it could also create conditions conducive to nutrient runoff from adjacent lands.

- Excessive nutrient loading could lead to eutrophic conditions and algal blooms, counteracting the intended restoration goals.

Risk of Unforeseen Ecological Responses

- The complex interactions within ecosystems make it challenging to predict all potential outcomes accurately.

- Unforeseen ecological responses might arise due to factors not accounted for in models and predictions.

Mitigation and Management Strategies

- To address potential negative consequences, adaptive management plans should include strategies for early detection, rapid response, and intervention.

- Monitoring invasive species, tracking species composition shifts, and understanding water dynamics are key components of effective management.

Collaboration with Experts and Stakeholders

- Engaging experts and stakeholders in risk assessment and decision-making processes enhances the ability to identify and address potential challenges.

- A collaborative approach helps foster resilience and adaptability in the face of uncertainties.

By critically evaluating potential negative consequences, this section acknowledges the intricacies of ecosystem dynamics and the importance of holistic understanding in restoration initiatives. Recognizing these challenges allows for the formulation of robust management strategies that can mitigate risks and enhance the success of the restoration effort. As we navigate the path of restoration, the awareness of potential negative consequences guides us towards a more informed and conscientious approach to the revitalization of the Salton Sea ecosystem.

7.2 Ecological Disruptions

Building on the potential negative consequences, this section will delve into the potential ecological disruptions that might arise from the intervention. It will explore scenarios in which changes in salinity and water quality could have cascading effects on species interactions, food webs, and overall ecosystem stability.

7.3 Overcoming Technical and Infrastructural Challenges

The successful implementation of the freshwater addition strategy, which includes the ambitious goal of producing 1 million liters of water daily from the air, demands a multifaceted approach to address technical and infrastructural challenges. This section delves into potential obstacles encompassing the intricate domains of designing and operating freshwater intake systems, transporting substantial volumes of water, and establishing the infrastructure needed for efficient water production, distribution, and ecological integration.

Designing Efficient Freshwater Intake and Production Systems

- The design of efficient freshwater intake and production systems must seamlessly integrate technologies to extract and generate water from the air.

- Balancing water quality, intake depth, and energy efficiency while preventing the entrapment of organisms becomes an intricate engineering feat.

Transporting and Distributing Water

- The transportation of substantial volumes of water, now including water generated from the air, necessitates robust conveyance systems.

- The infrastructure for water distribution must accommodate both the newly generated water and other water sources, ensuring optimal delivery to the Salton Sea.

Infrastructure for Water Generation and Distribution

- Establishing a comprehensive infrastructure for water generation from the air involves atmospheric water harvesting technologies.

- Integrating these systems with existing infrastructure requires careful planning to ensure seamless distribution and ecological integration.

Energy Considerations and Environmental Sustainability

- The process of water generation from the air necessitates energy inputs, sparking considerations for renewable energy sources.

- Minimizing environmental impacts and enhancing long-term sustainability through energy-efficient solutions is a central challenge.

Environmental and Ecological Integration

- The integration of newly generated water into the ecosystem demands consideration of water quality and ecological dynamics.

- Ensuring that the introduced water harmoniously integrates with the existing environment without causing disruption poses a significant challenge.

Regulatory Compliance and Stakeholder Engagement

- Overcoming technical and infrastructural challenges requires alignment with regulatory frameworks and engagement with stakeholders.

- Collaboration with regulatory agencies and local communities fosters transparency, addresses concerns, and ensures compliance.

Budgeting and Funding for Innovation

- The pursuit of water generation from the air introduces innovative technologies, necessitating budgeting for research, development, and implementation.

- Securing funding for cutting-edge solutions is instrumental for successfully surmounting technical hurdles.

Adaptive Management and Learning

- The complexity of technical challenges underscores the importance of adaptive management strategies.

- Flexibility in adapting to new technologies, responding to unexpected obstacles, and continuous learning are key components.

By ambitiously incorporating the production of 1 million liters of water daily from the air into the restoration strategy, this section underscores the need for interdisciplinary collaboration, innovative engineering, and a deep understanding of ecological integration. As we navigate the complexities of technology implementation and infrastructure development, the commitment to innovation guides us toward a sustainable future for the Salton Sea ecosystem, where technological advancement converges with ecological restoration.

Chapter 8: Policy and Management Recommendations

8.1 Developing a Comprehensive Restoration Plan

As the vision of introducing fresh water to the Salton Sea transforms into a tangible endeavor, the formulation of a comprehensive restoration plan emerges as a cornerstone for success. This section intricately lays out the key components that comprise this plan, illuminating the path forward with precision and purpose. From delineating specific goals and desired target salinity levels to outlining water distribution methodologies and establishing a timeline for implementation, each facet of the restoration plan converges to orchestrate a harmonious revitalization of the ecosystem.

Defining Clear Restoration Goals

- The restoration plan sets forth unequivocal goals that encapsulate the vision for the Salton Sea's transformation.

- Goals encompass ecological health, water quality enhancement, and the sustenance of vibrant habitats.

Setting Target Salinity Levels

- The establishment of target salinity levels serves as a benchmark for assessing the success of the restoration effort.

 These levels guide decision-making on freshwater input quantities and their distribution.

Designing Optimal Water Distribution Methods

- The restoration plan delves into the intricacies of water distribution mechanisms, encompassing canals, sluices, and natural flow patterns.

- Ensuring equitable dispersal of water across the lake is a vital aspect of the plan's execution.

Constructing a Realistic Timeline

- The restoration plan delineates a pragmatic timeline that accounts for various phases, from infrastructure development to water introduction.

- A well-structured timeline promotes efficient resource allocation and progress tracking.

Incorporating Adaptive Management

- The restoration plan acknowledges the dynamic nature of ecosystems and environmental conditions.

- Adaptive management principles allow for adjustments based on emerging data and unforeseen challenges.

Safeguarding Ecosystem Integrity

- The plan integrates strategies to protect and restore critical habitats, fostering the revival of native species.

- Incorporating measures to prevent disruptions to ecological dynamics is a priority.

Inclusive Stakeholder Engagement

- The restoration plan embraces the participation of stakeholders, including local communities, agencies, and experts.

- Involving stakeholders in decision-making fosters transparency, addresses concerns, and garners support.

Balancing Ecological and Economic Interests

- The plan navigates the delicate equilibrium between ecological restoration and economic interests.

- By considering economic benefits from enhanced ecosystems, the plan aligns restoration with sustainable development.

Continuous Monitoring and Reporting

- The restoration plan incorporates robust monitoring mechanisms to track progress and outcomes.

- Regular reporting enhances transparency and accountability throughout the implementation.

By meticulously weaving these components together, the comprehensive restoration plan emerges as a guiding compass, steering the course of action towards the revitalization of the Salton Sea. Through clear goals, adaptive strategies, and stakeholder engagement, this plan forges a path that honors the delicate balance between environmental rejuvenation and human well-being. As we embark on this journey, the restoration plan stands as a testament to our commitment to restoring the ecological vitality of the Salton Sea for generations to come.

8.2 Monitoring and Adaptive Management Strategies

In the intricate tapestry of the restoration effort, effective monitoring and adaptive management emerge as essential threads that weave together progress, insights, and course corrections. This section unfurls a comprehensive framework for monitoring, meticulously tracking the evolution of water quality, salinity levels, aquatic species abundance, and habitat health. Simultaneously, it introduces the dynamic concept of adaptive management, which harnesses real-time data and emerging insights to steer the course of action toward optimal outcomes.

Continuous Water Quality Monitoring

- Regular assessments of water quality parameters, including temperature, pH, dissolved oxygen, and nutrient levels, provide insights into the lake's health.

- High-frequency data collection allows for early detection of anomalies and informed decision-making.

Salinity Monitoring and Management

- Constant monitoring of salinity levels ensures adherence to target levels set in the restoration plan.

- Deviations from target salinity trigger adaptive adjustments in water distribution and introduction.

Abundance of Aquatic Species

- Monitoring shifts in the abundance and distribution of aquatic species elucidates ecological responses to restoration efforts.

- Monitoring extends to both native and non-native species, offering a comprehensive view of ecosystem dynamics.

Habitat Health and Biodiversity

- Habitat health assessments encompass the vitality of critical ecosystems and their support for native species.

- Biodiversity indices gauge the richness of species, aiding in the evaluation of restoration success.

Data Integration and Analysis

- The accumulation of diverse datasets necessitates robust analytical methods to derive meaningful insights.

- Data integration highlights correlations, trends, and potential cause-and-effect relationships.

Adaptive Management Principles

- Adaptive management embraces flexibility and responsiveness to real-time data and evolving conditions.

- Emerging insights trigger adjustments to restoration strategies, enhancing efficacy.

Iterative Decision-Making

- Data-driven decision-making drives a cyclical process of observation, analysis, adjustment, and evaluation.

- Each iteration enhances the restoration plan's precision and responsiveness.

Collaboration with Experts

- Engaging scientific experts and stakeholders enriches the interpretation of monitoring data.

- Collaboration ensures that insights translate into informed action.

Scenario Testing and Modeling

- Data-driven modeling simulates potential scenarios and their implications on the ecosystem.

- Scenario testing informs strategic decisions and prepares for varying outcomes.

Transparent Reporting and Communication

- Regular reporting of monitoring results fosters transparency and accountability.

- Open communication with stakeholders ensures collective ownership and support.

By intertwining continuous monitoring with adaptive management strategies, this section forges a path that is as dynamic as the ecosystem it seeks to restore. The iterative process of data collection, analysis, and adjustment empowers the restoration effort with resilience, adaptability, and precision. As we navigate uncharted waters, the commitment to rigorous monitoring and adaptive responses guides us toward a future where the

Salton Sea thrives in renewed vitality and ecological balance.

8.3 Engaging Stakeholders and Local Communities

In the orchestration of a restoration initiative as ambitious and transformative as the introduction of fresh water to the Salton Sea, the art of collaboration and engagement emerges as the linchpin of success. This section reverberates with the profound significance of involving an array of stakeholders and local communities in the very heart of the planning and decision-making processes. As we navigate the intricate dance of restoration, transparency, empathy, and shared ownership take center stage, ensuring that the journey toward renewal is a collective endeavor.

Inclusive Stakeholder Participation

- Stakeholders, ranging from government agencies and NGOs to local residents and environmental groups, are integral contributors.

- Diverse perspectives enrich the restoration plan, fostering holistic insights.

Transparent Communication

- Establishing transparent channels of communication instills trust and credibility.

- Sharing progress, challenges, and outcomes fosters informed engagement.

Listening and Addressing Concerns

- Actively listening to concerns and feedback from stakeholders reflects genuine consideration.

- Addressing concerns demonstrates a commitment to incorporating varied viewpoints.

Shared Vision and Ownership

- Fostering a sense of shared ownership ensures stakeholders are vested in the success of the restoration.

- Collective buy-in paves the way for smoother implementation and sustained support.

Collaborative Decision-Making

- Involving stakeholders in decision-making nurtures a sense of agency and influence.

- Collaboration ensures that decisions are collectively reasoned and well-informed.

Public Awareness and Education

- Educating the public about the restoration initiative enhances understanding and garners support.

- Awareness campaigns bridge knowledge gaps and promote active engagement.

Cultural and Social Considerations

- Recognizing cultural sensitivities and local values establishes respect and rapport.

- Incorporating cultural insights enriches restoration efforts and builds strong relationships.

Long-Term Relationship Building

- Engaging stakeholders in a sustained manner cultivates enduring partnerships.

- Long-term relationships promote a sense of investment beyond the project's completion.

Adaptive Engagement

- Adapting engagement strategies based on stakeholder feedback nurtures continuous dialogue.

- Flexibility ensures that engagement evolves alongside changing needs.

Transparency in Decision-Making

- Decisions related to the restoration are communicated openly, explaining the rationale behind them.

- Transparency builds trust and dispels misconceptions.

Through the interplay of engagement and collaboration, this section underscores that the journey toward restoring the Salton Sea is not only about ecological transformation but also about forging meaningful connections. As we navigate the realm of stakeholders and local communities, the commitment to shared values, open dialogue, and collective purpose propels us toward a future where the Salton Sea thrives as a testament to collective stewardship and the power of collaboration.

By integrating scientific insights with practical strategies for collaboration and adaptive management, this research will contribute to the development of a robust restoration plan that maximizes the potential benefits while mitigating risks and challenges.

Chapter 9: Conclusion

9.1 Summary of Findings

The exploration of the Salton Sea's restoration journey has led to a tapestry of findings that illuminate the potential for rejuvenation and renewal. Through empirical investigation, predictive modeling, and multidisciplinary analysis, this roadmap has unveiled insights that contribute to the understanding of the lake's complex dynamics and the possibilities for its ecological revival.

9.1.1 Hydrological Insights

The hydrological modeling conducted in Chapter 2 provided a comprehensive understanding of the lake's water balance, evaporation rates, and inflow patterns. It revealed the intricate interplay between natural forces and human interventions, offering a foundation upon which restoration strategies can be built. The introduction of fresh water, coupled with innovative technologies such as

Atmospheric Water Generation, emerges as a feasible approach that can mitigate water loss and reduce the reliance on unsustainable diversions.

9.1.2 Ecological Considerations

Chapter 3 delved into the potential ecological impacts of introducing fresh water to the Salton Sea. The assessment of biodiversity changes, habitat suitability, and aquatic species' response highlighted the transformative potential of the intervention. The roadmap underscores the importance of balancing ecological restoration with stakeholder engagement, emphasizing the interconnectedness of ecosystems and communities.

9.1.3 Economic and Societal Aspects

The examination of economic benefits and societal implications in Chapter 4 illuminated a promising horizon for local communities. The potential for increased tourism, enhanced recreational opportunities, and improved property values emerged as tangible benefits. However, the roadmap also recognized the need for careful planning to ensure that economic gains are coupled with the preservation of ecological integrity and social equity.

9.1.4 Navigating Challenges

Throughout the roadmap, the complexities and challenges of the Salton Sea's restoration journey were acknowledged. The potential for unintended consequences, ecological disruptions, and technical obstacles were discussed in Chapters 7 and 8. These challenges underscore the necessity of adaptive management, stakeholder engagement, and ongoing monitoring to ensure the resilience and success of restoration efforts.

9.1.5 A Vision for the Future

In conclusion, this roadmap presents a vision for the Salton Sea's future—one where the restoration of ecological balance intertwines with the aspirations of communities. It calls for a holistic approach that harnesses the power of science, technology, and collaboration to breathe life into the lake's waters. The roadmap's findings illuminate a path toward a reimagined Salton Sea—one that stands as a testament to the harmonious coexistence of nature and human stewardship.

As the final curtain falls on this exploration, the call to action resounds—an invitation for continued research, adaptive strategies, and the dedication of individuals and communities to realize the vision of a revitalized Salton Sea. The findings of this roadmap are not the conclusion of the lake's narrative, but rather a stepping stone toward its next chapter—a chapter marked by resilience, restoration, and a shared commitment to the legacy of this unique ecosystem.

9.2 Implications for Restoration Efforts

As the culmination of our research illuminates the potential of introducing fresh water to the Salton Sea, this section transcends the immediate context to cast a broader gaze upon ecosystem restoration efforts facing similar challenges in saline ecosystems worldwide. By synthesizing the findings, it unveils a tapestry of implications that resonate far beyond the shores of the Salton Sea, underscoring the significance of multidisciplinary approaches, technological innovations, and adaptive management strategies in addressing complex environmental degradation.

Multidisciplinary Approaches

- The research journey underscores the potency of combining ecological, hydrological, and socioeconomic insights for holistic restoration.

- Similar ecosystems can draw inspiration from this interdisciplinary synergy to develop comprehensive restoration plans.

Technological Innovations

- The integration of hydrological models, ecological simulations, and adaptive management techniques showcases the power of technology in restoration efforts.

- Advanced tools facilitate evidence-based decision-making and propel restoration toward sustainable success.

Adaptive Management Strategies

- The dynamic interplay between real-time data, monitoring, and adaptive adjustments serves as a paradigm for managing uncertainty.

- Embracing adaptive management as a guiding principle enriches restoration efforts' resilience and efficacy.

Broader Ecosystem Context

- Similar saline ecosystems facing degradation can draw parallels from the Salton Sea's narrative, fostering shared learning and innovative solutions.

- A collective approach to addressing ecosystem challenges fosters a global community of restoration practitioners.

Local and Global Relevance

- The implications resonate both at the local level of the Salton Sea and the global context of ecosystem restoration.

- Lessons learned and strategies employed transcend boundaries, contributing to a more sustainable planet.

Community Engagement and Stakeholder Collaboration

- The importance of involving local communities and stakeholders underscores the role of collective ownership in restoration success.

- Engaging stakeholders aligns restoration with societal values and aspirations.

Flexibility and Adaptability

- The research journey echoes the essence of flexibility and adaptability in response to evolving environmental dynamics.

- Restoration efforts must embrace change and harness adaptability as an inherent strength.

Inspiration for Future Restoration

- The research's implications serve as a beacon of inspiration for future restoration initiatives grappling with complex challenges.

- The Salton Sea's story becomes a source of hope, a testament to human ingenuity in the face of environmental adversity.

As this section draws the curtain on our exploration, it beckons the global community of restoration enthusiasts to reflect, innovate, and collaborate. The Salton Sea's potential renewal speaks to a collective commitment to restoring the harmony of nature. In the footsteps of this research, restoration endeavors worldwide are poised to elevate their pursuits, armed with knowledge, innovation, and an unwavering spirit to restore the beauty of Earth's ecosystems, one endeavor at a time.

9.3 Future Research Directions

As our journey through the potential restoration of the Salton Sea draws to a close, the path forward opens up to uncharted territories of discovery and inquiry. While this guide serves as a compass for understanding the benefits and challenges of introducing fresh water to the Salton Sea, it also lays the foundation for future explorations that hold promise in deepening our understanding and refining restoration strategies. This section casts a visionary spotlight on potential avenues for future research, igniting the spark of curiosity and innovation.

Interactions between Freshwater and Aquatic Species

- Investigate the intricate ecological interactions that ensue between the introduced freshwater and the existing aquatic species.

- Understand how varying salinity levels impact different species, trophic dynamics, and overall ecosystem stability.

Scalability of Atmospheric Water Generation

- Delve into the scalability of atmospheric water generation technology for producing large volumes of freshwater.

- Explore the feasibility of deploying such technology on a larger scale to address freshwater scarcity in other regions.

Long-Term Monitoring and Ecological Dynamics

- Embark on long-term monitoring endeavors to track the outcomes of the restoration initiative over extended time frames.

- Observe how the ecosystem responds and adapts to the introduced freshwater and gauges the restoration's long-term success.

Community Engagement and Socioeconomic Impact

- Investigate the socioeconomic impacts of the restoration on local communities, including changes in tourism, property values, and overall quality of life.

- Understand the role of community engagement in sustaining restoration efforts and fostering a sense of stewardship.

Ecosystem Services and Biodiversity Assessment

- Assess the restoration's impact on ecosystem services, such as improved water quality, enhanced carbon sequestration, and increased biodiversity.

- Quantify the tangible benefits that arise from the restoration, demonstrating its holistic value to society.

Innovations in Water Management and Conservation

- Investigate innovative water management and conservation practices that can complement the restoration efforts.

- Explore sustainable water use strategies to ensure the long-term success of the restoration and mitigate future challenges.

Global Application of Restoration Strategies

- Extend the insights garnered from the Salton Sea restoration to inform restoration strategies in similar saline ecosystems worldwide.

- Foster a global exchange of knowledge and best practices in tackling ecological degradation.

Policy and Governance Considerations

- Delve into the policy frameworks and governance structures that support large-scale restoration initiatives.

- Investigate how policy interventions can facilitate the successful implementation of similar restoration projects.

As we bid adieu to this chapter, we embark on a new chapter of exploration and discovery. The future research directions laid out here beckon passionate researchers, scientists, and enthusiasts to step onto the stage of innovation, with the Salton Sea as both a muse and a motivation. With every question pursued, every hypothesis considered, and every discovery made, the legacy of restoration grows, weaving a tapestry of renewal for ecosystems and a brighter future for our planet.

ABOUT AUTHOR

Antonios Valamontes is a visionary entrepreneur, seasoned investor, and trailblazer in digital publishing. With a profound passion for knowledge, he has dedicated his career to creating transformative educational content at the intersection of archaeology, history, and cutting-edge technology trends.

Recommended Other Books by Author

https://a.co/d/4UvEN3o   https://a.co/d/cjZhavJ

Sources

1. California Natural Resources Agency. (2021). Salton Sea Management Program. Retrieved from https://saltonsea.ca.gov/

2. United States Geological Survey. (2021). Salton Sea Data. Retrieved from https://waterdata.usgs.gov/nwis/inventory/?site_no=10294200

3. Audubon California. (2021). Salton Sea. Retrieved from https://ca.audubon.org/conservation/salton-sea

4. California Department of Fish and Wildlife. (2021). Salton Sea Ecosystem Restoration Program. Retrieved from https://wildlife.ca.gov/Conservation/Watersheds/Salton-Sea

5. Hoover, R. (2018). The Salton Sea: A Study in History and Politics. University of California Press.

6. Institute for Social Ecology. (2021). Salton Sea Restoration: Past, Present, and Future. Retrieved from https://social-ecology.org/wp/2006/05/salton-sea-restoration-past-present-and-future/

7. Hanak, E., Lund, J. R., Dinar, A., Gray, B., Howitt, R., Mount, J. F., ... & Medellín-Azuara, J. (2011). Managing California's water: From conflict to reconciliation. Public Policy Institute of California.

8. Imperial Irrigation District. (2021). Salton Sea History. Retrieved from https://www.iid.com/the-salton-sea-history

9. Faigin, D. (2015). Salton Sea. In California's Fading Wildflowers: Lost Legacy and Biological Invasions (pp. 7-24). University of California Press.

10. Richardson, L. A. (2005). Biodiversity and conservation of the Salton Sea ecosystem. Proceedings of the National Academy of Sciences, 102(30), 10871-10877.

11. Hurlbert, S. H., & Mulla, D. J. (2001). Once more on the dynamics of open acyclic communities. Ecology, 82(10), 2817-2832.

12. National Oceanic and Atmospheric Administration. (2021). Atmospheric Water Generator. Retrieved from https://oceantech.noaa.gov/research-and-programs/research-topics/atmospheric-water-generator

Image Sources

1. Photos by PublicDomainPictures. Retrieved from Pixabay.